PRAISE FOR WILD LOGGING

These stories about real people, living by a land ethic, can be a guide to life in the woods. The prose flows seamlessly between easygoing storytelling and sound, practical advice. Forest landowners throughout the country will be both entertained and enlightened.

—GREG APLET, forest ecologist at The Wilderness Society
and editor of *Defining Sustainable Forestry*

For the past several decades, forests and forestry in the U.S. have been the focus of seemingly endless controversy. My bookshelf is filled with polemics from one point of view and another, each claiming to represent the real truth. A well-worn expression in the American West is "whisky is for drinking, water is for fighting over." We seem insistent upon making forests, too, something to be fought over rather than enjoyed. Then comes Bryan Foster's Wild Logging. *In the modern literature of forestry, Foster's book is a fine single-malt scotch, to be cherished and savored to the last drop.*

Foster writes with a freshness and warmth, and unabashedly shares the genuine affection for forests and good land stewardship shown by the forest landowners who are the subject of his book. Missing is the stridency, and the "forest facts" that serve as thinly veiled surrogates for one activist agenda or another. In their place is a forthright description of Foster's odyssey across America, seeking to understand the nature of our varied forests and the equally varied men and women who know and love them like an old friend.

If one would seek to understand what forest stewardship really means, one need look no further than the people who occupy the pages of Foster's very readable, eminently practical, and warmly inspiring book.

—V. ALARIC SAMPLE, President, Pinchot Institute for Conservation

In Wild Logging, *Bryan Foster tells the stories of forest stewards who are motivated less by economic return than by their sense of allegiance to natural beauty and the larger community of life. By combining these examples with technical information about such matters as management plans and contracts, he encourages a more ecologically sustainable and culturally nourishing approach in the private, nonindustrial forestlands of the Intermountain West. This book is both a timely and a constructive one.*

—JOHN ELDER, Professor of English and Environmental Studies at
Middlebury College and editor of *The Norton Book of Nature Writing*

WILD LOGGING

*A Guide to Environmentally and
Economically Sustainable Forestry*

BRYAN FOSTER

With a Foreword by Jack Ward Thomas
Illustrated by Peggy Foster

With Special Thanks To
J. Munroe McNulty and Julie McNulty

2003
Mountain Press Publishing Company
Missoula, Montana

Printed with a soy-based ink on recycled paper.

All illustrations © 2003 by Peggy Foster
All photographs by the author unless otherwise credited

Cover photographs © 2003 Jeff Painter
Front Cover photos: *Ponderosa pine forests selectively
logged by Bill Potter, Montana woodlot owner*

Disclaimer
Landowners should contact professional foresters, financial consultants, and/
or attorneys before implementing any of the forestry practices described in this
book to ensure that the practices suit their property, meet their financial objectives, and follow federal, state, and local laws. The author and publisher, as well
as the landowners and professionals interviewed here, waive all liability for
practices described in this book.

Library of Congress Cataloging-in-Publication Data

Foster, Bryan C., 1970–
    Wild Logging : a guide to environmentally and economically sustainable
forestry / Bryan Foster ; with a foreword by Jack Ward Thomas ; illustrated
by Peggy Foster.
        p.   cm.
Includes bibliographical references and index.
    ISBN 0-87842-448-2 (alk. paper)
    1. Sustainable forestry—West (U.S.) 2. Sustainable forestry. I. Title.
    SD144.A18 F67 2002
    634.9′2—dc21
                                                            2002013679

PRINTED IN THE UNITED STATES OF AMERICA

MOUNTAIN PRESS PUBLISHING COMPANY
P.O. Box 2399
Missoula, Montana 59806
406-728-1900

*This book is dedicated to
the memory of Joseph Munroe McNulty*

CONTENTS

ACKNOWLEDGMENTS

Foremost, I thank Joseph Munroe (Mun) McNulty, the philanthropist who funded this project. It was through his vision that I started this project and his patience and generosity that I finished it. He and his wife Julie made real my dream of writing a book.

I also thank: Jack Ward Thomas, former chief of the U.S. Forest Service, who gave detailed comments to guide this book from its most crude versions to its most refined; my family, Boyd, Peggy, and Paul Foster, for their support; Katherine Cooper for her encouragement; the landowners I feature in this book, for it is their work which can inspire others (Merv and Anne Wilkinson, Leo and Marilyn Goebel, Bob and Inez Love, Bill and Vicki Byrd-Rinick, David and Kathryn Owen, Bill and Betty Potter, and Mel and Betty Ames); everyone who offered comments on a draft of this book (Gregory Aplet, Paul Barton, Gary Ellington, Nancy Fresco, John Gordon, Fred Hodgeboom, Keville Larson, and Ed Marston); all other forest owners and professionals I interviewed, for they all shaped this book whether or not they are directly mentioned in the text (Nathan and Steve Arno, Steve Bennett, Clinton Bentz, John and the late Shirley Bowdish, Hal and Arlene Braun, Sam Brown, Scott Burnham, Bill Butler, Orville Camp, Al Cota and Tom Elliott, Jim Costamagna, Howard Dixon, Gary Ellington, Larry Evans, Charlie Fitzgerald, Jerry Franklin, Marsha Goetting, Hank Goetz, Herb Hammond, Patrick Heffernan, Rudy Hickson, Fred Hodgeboom, Mitch Lansky, Chuck and Rose Lane Leavell, Brooks Mills, Chad Oliver, Art Partridge, J.R. Reynolds, Frank and Karen Sherwood, Carol Ann Wassamuth, and Harris Wiltzen); Jackson Hole Mountain Resort for work and play while I was writing the book; and the whole team at Mountain Press, particularly my editor, Lynn Purl, for her insight and forbearance.

The first step is to throw your weight around on matters of right and wrong in land use.

— Conservationist Aldo Leopold
in *A Sand County Almanac*

The days have ended when the forest can be viewed only as trees and trees only as timber.

— Senator Hubert Humphrey, from a
1976 speech to the United States Senate

FOREWORD

Forests in the United States fall into three categories—federal and state government, private industrial, and private nonindustrial. Within that third category lies the land managed, and treasured, by the people featured in this book. Bryan Foster has focused on landowners with a vision for the future of their land and a true sense of stewardship in their hearts.

Each of these landowners managed his forest in different circumstances of landforms, weather, and ecological relationships. Yet discernible themes run through each of these stories—a deep love of the land, a strong work ethic, and a vision of what can and should be. These landowners seem to have a full understanding of what Aldo Leopold meant when he talked of forest managers "writing their signatures on the land" and "being fully conscious while chopping." Each landowner was determined to write his signature carefully and proudly so the next generation of stewards would know his work and inherit a worthy legacy. It is through these stories that we appreciate what can happen through vision, hard work, and a land ethic.

I was impressed with Foster's storytelling ability. I almost felt as though I were with him as he visited the various forest stewards featured in this book. This book is a well-woven tapestry of people and places, of hard work and dreams. It is not a dull rendition of facts, figures, and mathematical equations. Foster features strong characters with a story to tell and a dream to share—and then makes the places, and most certainly the characters, come alive.

This book is not a cookbook for practicing forestry. In fact, the book clearly demonstrates that forestry involves different tools and approaches for achieving a landowner's objectives. Those objectives are clearly circumscribed by the current and potential condition of the forest. But each landowner developed a vision of what can become of his forestland, even after the land has passed on to others.

This book is a place to start for owners of small forest tracts. Its clear purpose is to encourage other landowners to become superior stewards of the land. Foster provides a means to such a vision and enough information to start the journey.

So, read on. There is inspiration here—and some very good writing. Build a dream, that most wonderful attribute of the mind and heart. Then you might be inspired to begin your own journey of informed forest stewardship.

—JACK WARD THOMAS
Boone and Crockett Professor of Wildlife
Conservation, University of Montana
Chief Emeritus, U.S. Forest Service

Introduction

COMING INTO THE WEST

[This is] a country of unparalleled fertility, salubrious climate, rapid growing seasons, railway transportation and electric light in the near future. To those who love flowers, trees and birds this is a wonderland. The mud here isn't sticky and the water is soft.

—THE ANACONDA COMPANY, FROM *THE CALL OF THE LAST WEST,*
A 1938 BROCHURE ADVERTISING LAND IN THE WEST

Badger-Two Medicine mountains in northern Montana

THE DRY, BURLY REGION from eastern Oregon and Washington through Idaho to western Montana has variously been called the Intermountain West, the Inland West, the Northern Rocky Mountains, and the Inner-Columbia River Basin. For some, the West extends from the 100th meridian west to the Pacific Ocean (running from North Dakota south through Texas, the 100th meridian is the approximate line where the lush east gives way to the arid west, and the average annual precipitation drops below 20 inches). But for me, only the intermountain region is the genuine West. Classic western books such as *The Big Rock Candy Mountain* and *The Last Best Place* come from here, as do the cowboy and Indian paintings of Charles M. Russell. Farming, mining, and logging towns still polka-dot this region's highways, and only the city of Spokane has a population above 150,000. Here lies the largest stretch of unbroken forestland in the lower forty-eight states — more than 30 million acres. In parts of it, grizzly, wolf, and lynx still live.

I grew up in Montana, and after working long days on my father's farm, I calmed myself most evenings by walking in the woods. Sometimes I would climb up on the farm trees, pretending I was in a Swiss Family Robinson perch. Other times, I would just lie beneath them, listening to cracks, rills, and chatters 50 or 100 feet up. Standing beneath a ponderosa pine, I tore off pieces of bark and tried to puzzle them back together. In the nearby Beartooth Mountains, I carved my initials into a thick Douglas-fir, astonished and exhilarated at the permanence of the pocketknife mark. I became enthralled with trees' ages, and their tall, cool shade, during those summers on the farm.

What would I do if I inherited this place? I asked then. What would I do with this burden and gift of land?

Years later my father sold the farm. I decided to go to graduate school in forestry anyway, on the chance that someday I'd own forestland. I visited various forests while I was in school: groves of thousand-ton sequoia trees in California, their tops lost in cloud and vapor; deciduous forests in Connecticut filled with trees of flavorful names like elm and maple, hickory and cherry; Arkansas pine plantations that pumped out wood every thirty years; a linear stand of spruce in Germany, enlivened with a patch of beech to reduce soil acidity from the conifers; and a rainforest in Puerto Rico crowded with more than two dozen tree species on each acre, trees with names such as *Cecropia, Dacryodes,* and *Miconia.*

While I was still in graduate school, I met Mun McNulty, a Yale University alumnus and an environmental philanthropist. Whenever McNulty wasn't running his family's construction company in Chicago, he was out fishing, hiking, hunting, or skiing in the West. He had visited a guest ranch in Montana many times over the years. There he became interested in the fact that one of the owners, Bill Potter, harvested young ponderosa pine while leaving many of the older pines.

Bill Potter and one of the yellow-belly ponderosa pines on his E-L Ranch in Montana

The E-L Ranch has been in the Potter family for three generations. Today, Potter runs a guest ranch and harvests timber on 2,000 acres along the Blackfoot River near Missoula. Vacationers come to the E-L as much for the grandeur of the ponderosa as for the gourmet meals, horseback rides, and trout fishing.

McNulty wanted to see more land managed as elegantly as Potter's property. He was concerned that greater numbers of people would soon be buying forestland in the Intermountain West, an increasingly popular region with growth rates of 3 to 13 percent per year (compared to 1 percent for the country as a whole). He wanted to make sure that the new landowners managed their forestlands well. If they were knowledgeable about western forest ecology, timber harvesting methods, and the economic issues surrounding their forest estate, they wouldn't easily be taken by slick loggers or developers. McNulty envisioned a book that would present technical information to owners of forests alongside stories from successful landowners, demonstrating forestry practices learned through experience.

I met McNulty on the wooden front porch of one of the E–L cabins late one summer to talk about his idea. Over drinks he offered me the opportunity to research and write this book.

I started my search for landowners in 1998 by asking for recommendations from the American Tree Farm System. With about 70,000 members, it is the oldest and largest landowner organization in the country. To join you must own at least 10 acres of forestland and meet standards for forest management, judged by an independent forester (see Appendix I for contact information). Every year the association's committee of professional foresters chooses an exemplary tree farmer from each state and then from each region.

I went on the road for two months and visited thirty-five landowners and forestry professionals, including landowners from the recent Tree Farmer of the Year list and others I had read about or been referred to.

To get to the landowners, I drove through dozens of small towns, scatterings of old houses and trailers. Some towns had been named by students from eastern colleges who worked summers on the railroad — Princeton, Dartmouth, Purdue — and others were named for favorite prostitutes — Juliaetta, Deary. In eastern Idaho, among patches of cedar and blooming cherry, the Nez Perce gave towns mellifluous names — Kootenai, Lochsa, Mescale.

Sometimes I stopped at these towns. I played pool one night with a man who packed a chew as big as a fist. Another time, I stopped for a lunch of grilled cheese and apple pie in eastern Washington. A truck with the bumper sticker "If you can't mine it, you gotta cut it!" idled outside the whole time I was eating.

The journey between the landowners' properties was long. Each morning, when I started driving fresh, the scenery of talon peaks and flat-bread plains was exhilarating. But after a full day of driving, the road became a greasy set of switchbacks or a dull, thudding straightway. During one drive up a mountain pass in Oregon late at night, I fell asleep (I was crossing the breadth of the state from Selma in the southwest to Joseph in the northeast). When I shook myself awake, I was at the top of the pass, with my truck mysteriously parked to the side of the road and the engine turned off. Another time, I traveled from rural Wyoming to rural Maine, spending eight hours on each of two days in cars, trains, subways, and planes.

Finally, as I got close to each landowner's property, I checked my scribbled directions and the mileage on the odometer against the landmarks I passed — curves, forks, railroad tracks, mailboxes. My truck would buck over muddy roads, stagger through stretches of deep gravel, and edge along hillsides. After I arrived, the landowners and I might talk in the living room around stacks of books about camping and forestry, gardening and cooking, then walk into the woods. Other times we piled firewood, cut brush, or thinned out small trees, and talked later.

I asked dozens of questions, but mainly wondered how these people took care of their trees. Most said they sought training and professional advice from extension services and state and private forestry consultants. Some eventually tried to manage the land themselves. All of them, whether working by themselves or with others, found out through trial and error in the forest which practices work and which don't.

One of the first landowners I met in the West was J. R. Reynolds, an insurance salesman and judo artist who lives on 44 acres of land near Missoula, Montana. I was looking at photos on the wall of a coffee shop in Missoula when I saw him nearby talking with a friend. His eyes were bright, his hair was tinged with gray, and he wore a green Tree Farm cap. I recognized him from a conference for forest owners I'd attended a week earlier, so I stopped him on his way out of the coffee shop. He pumped my hand after I introduced myself, and we sat down on a bench.

"Do you own land around here?" I asked.

"Yes," he said. "It's a beautiful place." He told me about rattling down a hill in front of his house on a pair of old Kneissl White Star skis in the winter—he would build small jumps at the bottom of the hill. He talked of walking the trails in the summer, measuring the season by the sequence of wildflower blooms: trillium, glacier lily, yellow violet, shooting star, caliope, ninebark, mock orange, snowberry.

"My land's just up that valley." He pointed toward the mountains. "Come out anytime." He gave me his phone number just before leaving.

As soon as I arrived at Reynolds's place that winter, we put on skis and went out into the forest. We crunched and skidded through the snow. We glided up and down hills and sped across flat meadows. Then, after a half hour of skiing, we stopped, each of us panting, near a giant western larch, a rare specimen over two hundred years old.

"Where are you from originally?" he asked, his words jagged from heavy breathing.

"Great Falls," I said.

Everything was silent then, except for our breath, as heavy as horses', and the late winter sounds of snow settling and melting.

"What about you?" I asked.

He told me he was from Connecticut but that he and his first wife decided decades ago to move west. They spent eight months driving around the Northwest, looking for somewhere to live. This place seared into them, so that wherever they went, they kept thinking of Missoula. "There's nothing else like this," he said. "There are those tall Bitterroot Mountains and that big Bitterroot River."

"So what about your first wife?" I asked.

Reynolds told me he had lost two wives because of this land—the first left because she preferred socializing in the city over doing chores in the country, and the second left because she wanted to walk to work instead of driving an hour each way (she has since moved back). Reynolds admitted that there were other problems in these marriages, but they seemed less important to him than the land. "If anyone is considering buying forestland, I would tell them that the decision needs to be shared between the spouses," Reynolds said, adding that the remoteness of forestland and the time for managing it will only stretch apart a weak relationship.

Reynolds took off again, skiing fast through the trees, his ski cap shaking with pins from the National Ski Patrol, American Red Cross, Alta Ski Resort, and Mutual of Omaha Insurance Company. His arms swung as he pointed things out to me.

"That's my big ol' larch," he said. "That's the saggy duck pond." I visualized the soil as a brown cloth, sagging down where water pooled. There were a few deciduous trees here, a few cottonwood and alder by the pond. These trees can afford to rebuild their leaves every year because the pond water buffers them from the dry winters and summers.

"It was important for me to do a title search and boundary survey when I first bought this property so I knew whether these fences really marked my land," said Reynolds, in one quick breath, when I caught up to him by a fence post. Then he turned and sped on.

After another long stretch of kicking and gliding, we arrived back at the house and took off our skis. My muscles had adjusted to the motion of skiing, and it took a few minutes for me to be able to walk smoothly. We put our skis away, the cold air sharpening the colors of a winter sunset.

"I'm constantly amazed by this place," Reynolds said. "But then I don't know enough to be able to do much, except maybe cut a few of the dead trees into firewood. I mean, everything I do seems to have some impact. If I feed chickadees and nuthatches, sharp-shinned hawks and merlins make this their flyway; if I bring ducks into the pond, ermine come down to get them. How can I do anything with these trees when I know I might affect the soil biota or change bird nesting or something else? I'd rather not cut timber here. I'd rather just leave this place as a refuge, a sanctuary."

About half the landowners in the Rockies are like Reynolds, worried that cutting timber would ruin their land. Some of them protect their land during their lives, but their spouses or children later have to subdivide the property or cut its timber to pay estate taxes (see Chapter 3 for information about estate management). Other landowners harvest timber but turn everything over to the loggers and find their forest has been left weak and withered (see Chapter 2 for information about logging contracts). After visiting Reynolds, I sought out landowners who carefully harvest to boost the value of timber on their property, to calm disease infestations, or to reduce wildfire hazard, while still maintaining the purity of their water, the integrity of their soil, and the diversity of their wildlife. I was looking for model landowners.

During my travels, I visited Merv and Anne Wilkinson, who own forestland in British Columbia. I met Merv when he was 85. He still manages his land, with hands crooked from arthritis and always shaped, he jokes, for a chainsaw or ax handle.

Wilkinson describes his 136-acre "Wildwood" property as a "sustained-yield, selectively logged tract." He doesn't cut more wood than the forest grows, he leaves dead wood on the ground to add organic material to the soil, and he removes crooked, stunted, or dying "junk" trees around his big fir. He has harvested wood from it since 1945, keeping within the growth rate of 2 percent per year. He avoids high transportation costs and low mill prices by milling his wood on site with a portable Wood-Mizer bandsaw.

In August 1993, the Wilkinsons joined environmental activists in a blockade to stop MacMillan Bloedel from clearcutting a forest on Vancouver Island. After his arrest, Merv told the judge, "My Lord, it is not necessary to destroy the forest to extract timber. It is a matter of method." The Ecoforestry Institute, which will eventually manage the Wilkinsons' land, promises to uphold this philosophy.

Merv Wilkinson leaves the biggest, straightest fir to grow as much as possible and cuts away the "junk" around his crop trees.

To get to the Wilkinsons' place on Vancouver Island, I drove around a small lake along a road as smooth as taffy up to their stone and timber cottage. "Come on in and have a seat," Merv said as I met him at the front door. He encouraged a small fire in a wood-burning cookstove. His wife, Anne, came down from upstairs and started fixing us coffee. She is a quiet woman—small, with bright eyes and a nest of curly white hair.

Anne leaned over me at the kitchen table, set down a cup of coffee, and said softly, almost whispering in my ear, "Do you see that tree over there, that one with the dead top?" She pointed out a sliding-glass door to the forest across the lake.

"Yes," I said quietly. The tree stood about 50 feet above the surrounding trees, as if it had survived a fire or windstorm that had knocked down everything else around it.

"We call that tree Othello," Anne continued, standing up but still speaking softly. "It's an old warrior, probably 1,500 years old, and it's been through a lot in its life. It was hit hard by lightning about six years ago. We fed it a mixture of sawdust and chicken manure that year as a Christmas present." She looked at the tree and nodded.

Merv Wilkinson leaves some windthrown trees to deposit nutrients back into the soil.

The Wilkinsons, like all of the landowners in this book, love their forest-land. One of the landowners I visited has tied up a crooked branch of a be-loved larch tree to try to straighten it out. I have a photo of another landowner, gray haired and flannel shirted, putting his hands on the hips of his favorite maple and kissing it. All these landowners fix up little parts of their property like teenagers buffing up the chrome of their cars.

Eventually I chose the people to profile in depth for this book: landowner Leo Goebel near Joseph, Oregon; logger-forester Bob Love near Whitefish, Montana; and landowners David and Kathryn Owen near Seeley Lake, Montana. I chose them based on their willingness to share their knowledge and their lives with me and on their ability to manage their land in a fashion unlike any other I had seen. Their strategies for managing forestland were innovative—maintaining diverse tree species and multistoried forest structure—and financially sound. They all reap net profits from logging of $100 to $150 per acre, although Bob Love's clients let him reinvest these profits in precommercial thinning. Not all landowners have the time, money, or training to manage their forests as these owners do, but these are models to inspire others.

In this book, landowners tell what they have learned about managing their land, from completing forest inventories to choosing loggers. Chapter 1 profiles Leo Goebel and describes methods for inventorying forestland and determining a sustainable harvesting volume based on average annual tree growth. Chapter 2 shows logging close-up with Bob Love, who combines forestry (the thoughtful work of measuring trees and developing management plans) with logging (the sweaty work of felling and moving timber). This chapter describes the process of choosing a logger and lists the common elements of sustainable forestry, such as reduced road densities and restrictions on harvesting around streams. Chapter 3 shows David and Kathryn Owen converting a logged-over forest into a forest capable of growing commercial timber. This chapter focuses on estate management, as the Owens recently wrote a conservation easement to protect part of their forestland from future housing development. The epilogue features landowners applying the principles of sustainable forest management in another part of the country, the Maine woods. Appendixes include a list of organizations that can be contacted for advice, a sample timber harvest contract, and information on preparing a conservation easement.

The forests featured here represent many of the forest types in the region, and sidebars discuss management for other major forest types, such as prescribed burning for ponderosa pine/western larch forests.

This book is not for everyone. It is for people interested in forest management that is innovative, ecological, sustainable, and wild. None of these logging methods maximizes profits, but all are profitable; in particular, they are suitable for many landowners who have other income from urban careers, farm/ranch work, investments, or retirement savings.

This book deals almost exclusively with uneven-aged logging practices, where trees are selectively cut, leaving behind trees of differing ages (see the Technical Note "What Forestry Terms Do I Need to Know?" on page 14 for a more thorough description of uneven-aged logging). When done right, uneven-aged logging retains the forested beauty and wildlife habitat that landowners say they value most from the land. Uneven-aged logging also mimics the most frequent natural disturbances in the region—small patches of trees dying from fungal infestations or small burns, a single tree dying from a blowdown or insect attack. The drawbacks are that uneven-aged logging is generally more expensive and labor intensive than even-aged logging (clearcutting and other techniques where all trees of the same age or size are taken). Because uneven-aged logging involves more frequent harvesting, can impede regeneration of shade-intolerant species, and leaves standing trees more vulnerable to blowdown or disease infestation, it must be done carefully.

Even-aged, clearcut-type logging may be appropriate in some instances: in highly shade-intolerant forest types, such as lodgepole pine; when the woodland owner is trying to maximize his economic return; or in forests that have so lost their natural species and structure, whether through logging, planting, or other disturbances, that it is easiest to start over. But even-aged harvests rarely mimic natural disturbances in the West. Severe disturbances such as crown fires and windstorms are two to three times less frequent than the sixty-year harvest rotations typically employed by timber companies in the region. Also, clearcutting leaves areas shorn, unlike natural disturbances, which leave forests stubbly, with patches of dead and live trees.

This is primarily a book for landowners and managers who wish to harvest timber for profit, while also maintaining their forests' biological diversity, aesthetic beauty, and ecological structure. But even though this book focuses on landowners in the Rockies, it is applicable to managers of other land types in other regions. For example, a number of private companies, such as Anderson-Tulley in Arkansas, Seven Islands in Maine, and Collins Pine in Oregon and California, use the uneven-aged management practices promoted in this book. Government land managers Jensen Bissell at Baxter State Park in Maine and Gary McCausland at Fort Lewis Army Base in Washington also use uneven-aged management. The Forest Stewards Guild, an association of foresters committed to ecologically responsible forestry, keeps a database of land managers who practice innovative forestry across the country (see Appendix I for contact information).

Although landowners should read and learn as much as they can about their forestland, they should not try to manage the land without professional help, including foresters, loggers, lawyers, and accountants.

In the Nez Perce National Forest in Idaho, Earth First! activists and loggers have been fighting a stereotypical battle for almost a decade. The loggers will never understand the activists—their hot, young blood, their feeling that forests are something to be saved like a poem. Nor will the activists understand the loggers—their happiness at working in the forest, their urgency about wanting to keep jobs they have built families and lives around.

The landowners in this book have found a middle way between the airy idealism of environmentalists and the steel-toed practicality of loggers. These landowners want handsome forests above all, and have developed ways to take out some of the weaker trees to make a stronger forest for the future.

What do these landowners mean by "wild logging" or "sustainable forestry"? One of the oldest sustainable forestry concepts is Dauerwald, from the German *dauer,* meaning permanent, perpetual, or continuous, and *wald,* forest. In 1922 German forestry professor Alfred Möller developed this management philosophy, which calls for abstaining from clearcutting and substituting natural regeneration for tree planting.

More recently, Jerry Franklin, a forest ecologist at the University of Washington, has defined sustainable forestry as "removing timber while maintaining the integrity of the soil and water, and sustaining plant and animal life. The idea is to manage for a group of organisms and processes that create and maintain the forest, rather than just managing for trees and ignoring the structure that enabled those trees to grow."

Jack Ward Thomas, former chief of the U.S. Forest Service, agrees that sustainable forestry considers the forest as an ecological system and not simply as a timber resource. Sustainable forestry also implies a broader concept of ecosystem management, according to Thomas, one that considers both human needs and ecosystem capacities, and is farsighted as it looks at the effects of forest management beyond property boundaries and beyond the next few years.

In general, policy makers define sustainable forestry as providing for the needs of both current and future generations. But it's hard enough to think a decade ahead, much less a generation. Scientists often define sustainable forestry as maintaining the forest's diversity of organisms, its functions of nutrient, water, and gas cycling, and its vegetative structure. But it's difficult for small landowners to accurately measure these factors.

There are no firm guidelines for sustainable forestry, other than aiming to restore or maintain a natural forest. In areas thick with invasive plants and dark from tree shade, spraying herbicides and planting seedlings may be necessary. In wet, productive forests, some larger trees might be taken to bring in economic returns and improve growing space for trees lower down in the canopy. This book offers rough guidelines for sustainable forestry, gathered from the wild logging of innovative western landowners.

The Institute for Sustainable Forestry, a California-based nonprofit organization, has established a practical set of guidelines for sustainable forestry. The following "Ten Commandments" were inspired by these guidelines, and the landowners' practices described in this book. These guidelines call on managers to maintain the following forest attributes:

1. timber volume, harvesting less than the overall growth rate of the forest

2. aesthetics, ensuring that the forest is still attractive after the harvest

3. structure, keeping the same number of canopy stories and amount of standing and fallen dead wood

4. natural disturbances, reintroducing natural processes such as fire

5. natural regeneration, keeping enough trees to allow them to reseed naturally

6. soil fertility, reducing canopy openings and ground scarification

7. water quality, reducing chemical applications and the number of roads and intensity of timber harvest around water

8. unique cultural sites, protecting Native American ruins or old homesteaders' cabins

9. native species populations, maintaining the presettlement diversity of native plants and animals

10. old growth, protecting areas where more than half of the trees are greater than 200 years old

How do you implement sustainable forestry methods? First, as a landowner you should develop objectives for what you would like to do with your property, such as bringing in enough income to pay for property taxes, leaving a better forest for your children, or reducing wildfire and disease hazards. Second, you or a forestry consultant should inventory the timberland. Then you can look at the inventory and objectives together and decide whether or not to harvest timber.

If you decide to harvest, you or the forester must set conditions on the loggers for the harvest, following guidelines such as the Ten Commandments. You might designate streamsides and old-growth areas off-limits to timber harvesting. Growth rates established during the inventory might establish a ceiling for harvested timber volume. You could decide to save some big, standing dead trees, or "snags," for wildlife habitat. You might choose to harvest dying trees preferentially over their healthy neighbors. Roads can include a series of "dips and strainers," structures that divert erosive runoff. You might schedule logging in wintertime, when snow cover protects plants from machinery tracks and frozen soil resists compacting.

All this planning requires imagination—the ability to picture what the forest will be like after the harvest. Imagination combines information—such as inventory data and management options from state extension or U.S. Forest Service publications—with intuition—what looks and feels right, what respects other living things and maintains natural processes. This intuition is akin to the "land ethic" that conservationist Aldo Leopold advocated in *A Sand County Almanac,* where people give the land the same respect they give other people— a kind of Golden Rule for the forests. Tom Brown Jr., author of *The Tracker* and numerous field guides to plants and animals, similarly calls on landowners to manage their land not so much by the numbers as by their conscience, or what he calls "inner vision."

To develop this intuitive sense, you need to spend time in the forest. Breathe the air from the trees so often you smell like pine, feel the soil so regularly it becomes part of your skin, live on the land through some moody seasons. Then start to look at the forest. Focus on both the small and large scales—look for different species and ages of trees and trees that are dead; identify other plant and animal species, such as yellow chanterelles, horsehair lichen, pine marten, and mountain salamanders. Compare managed and unmanaged forests. Begin to see the big picture—trees becoming forests, forests becoming forest types, forest types becoming watersheds, watersheds becoming landscapes.

Intuition can mislead you, and information can be faulty or incomplete. The key is to examine and reexamine decisions, try things differently if they don't work the first time, talk to other landowners and forestry experts to gain their perspectives, and continue to spend time in the forest to understand its processes. Now listen to these landowners and what they have learned from the land.

TECHNICAL NOTE

What's an NIPF?

A Nonindustrial Private Forest is a woodland capable of producing commercial quantities of timber, but without timber processing facilities such as saw or paper mills. NIPF owners include individuals, couples, families, and friends. About 5 percent of NIPF acreage is owned by nontimber businesses such as pension funds, foundations, university endowments, and American Indian reservations.

NIPF Owners Are:

▶ **Important.** Most NIPF owners hold less than one quarter section (160 acres) of forestland, but some hold thousands of acres. Currently, NIPF owners hold between one-quarter and one-third of the total forestland in the Intermountain West and supply as much of the region's lumber.

▶ **Flexible.** NIPF owners, such as farmers and ranchers or retired lawyers and doctors, typically have limited forestry knowledge and equipment compared to industry and government personnel. But they can harvest whichever trees they choose (within the law) because they are not governed by high mill demands or strict agency rules.

▶ **Environmentally conscious.** Most NIPF owners say they own their land not to harvest timber but to enjoy the view, protect the wildlife, and pass along something tangible to their children. NIPF owners generally favor amenity values over commodity production and do not depend on their land as a primary source of income.

▶ **In need of help.** NIPF owners are often well educated, but they rarely have forest management plans. Landowners in the Intermountain West in particular do not have the information and technical assistance provided to landowners in other parts of the country, such as the Northeast or Southeast, where NIPF lands constitute over half of the timber base. Michael Dombeck, former chief of the U.S. Forest Service, said at a 1996 symposium that the country must "dramatically increase the number of NIPF owners who complete forest stewardship plans over the next century."

 TECHNICAL NOTE

What Forestry Terms Do I Need to Know?

You don't need to be a forestry professional to read this book, but it's helpful to become familiar with some of the technical terms used by foresters and loggers. Some of the most important terms are explained below. Most others can be found in the glossary.

▶ **Basal area** is the cross-sectional area of a tree at breast height, 4½ feet off the ground. Basically, it is the surface area you would see from a plane looking down at a forest if all the trees had been cut off at breast height. Basal area is a common measure of forest density or stocking, and also accounts for tree size (unlike trees per acre, the other common measure of density). A measurement of 30 square feet of basal area on 1 square acre (44,100 square feet) means that, for example, about thirty trees of 14-inch diameter or sixty trees of 7-inch diameter could cover the land. Foresters generally recommend harvesting ponderosa pine and western larch forests if their basal areas exceed 100 square feet per acre, as this indicates that the trees are crowding each other out—restricting timber growth and increasing the stand's susceptibility to fire, insects, or disease. Higher elevation grand fir and Engelmann spruce, which grow in a wetter habitat, are commonly harvested only when their basal areas exceed 200 square feet per acre.

You can assess basal area with a cruiser's crutch or prism—tools available from a forestry supply company.

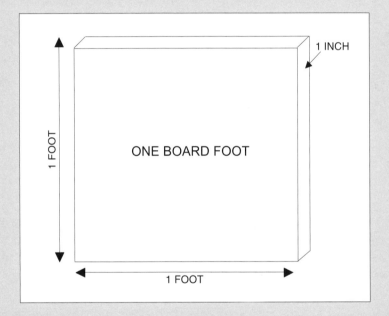

▶ **Board foot** is a common unit of timber volume. One board foot is the equivalent of a piece of wood 1 foot long by 1 foot wide by 1 inch thick. You'll commonly see the abbreviations "bf" for board feet, "mbf" for thousand board feet, and "mmbf" for million board feet. Landowners are paid for timber on the basis of board foot volume for different species—often a stumpage fee, paid by loggers to the landowners before cutting timber.

A trained forester can quickly estimate the volume of merchantable wood in a tree. In a few seconds, he can lay trees down in his head so they're cut up and stacked into boards. Although you may never be able to visualize wood like this, you can still learn how to measure the timber on your property by measuring the diameter of a tree with a tape measure and height with a clinometer, a special instrument that measures angles, available from forestry supply companies. A table, specific by species and available at state forestry offices, can then convert height and diameter into board-foot volume. A board-foot scale called "Scribner" is most commonly used in the Intermountain West.

These approximations will give you a better idea of board foot equivalents:

- 1 cubic foot equals 4 to 6 bf

- 1 merchantable tree in the Intermountain West typically contains 75 to 750 bf

- 1 ton roughly equals 250 bf

- 1 firewood or paper pulp cord roughly equals 2 mbf

- 1 truckload roughly equals 5 mbf

- 1 average three-bedroom house takes 15 mbf or 3 truck-loads

▶ **Dbh** stands for diameter at breast height. The tree's circumference is measured on the uphill side of a tree, 4½ feet above ground level. Trees 4 to 6 inches dbh typically sell at low prices for firewood or paper pulp. Trees 6 to 10 inches dbh sell at slightly higher prices for fence post and pole material. Trees 10 to 16 inches dbh sell at moderate prices for dimensional lumber. Trees greater than 16 inches dbh sell at handsome prices for plywood or furniture veneer. Timber prices depend not only on size but also on tree species (ponderosa and white pine are two of the West's most valuable conifers) and wood quality (including taper and straightness, number and tightness of knots, and number and extent of defects). You can measure dbh with a logger's tape, which has a scale for converting circumference into diameter. (With a regular measuring tape, take the circumference and divide by 3.14 to get the diameter.)

▶ **Even-aged logging practices** include clearcutting, seed tree harvesting, and shelterwood harvesting. In clearcutting, loggers harvest all the trees in a stand at one time. In seed-tree harvesting, they cut all the trees except for a few scattered and sturdy individuals that provide seed for the next generation. In shelterwood harvesting, loggers take half the trees; five to twenty years later, after the standing trees have provided shelter for seedlings, the sheltering trees are cut. The new forest that regenerates after any of these practices will have trees all about the same age. Even-aged and uneven-aged logging or forestry practices refer to a complete management regime of periodic harvesting.

▶ **A stand** is a forested area with sufficiently uniform tree species, forest stories, and average tree age to be distinguishable from the rest of the forest.

▶ **Thinning** refers to logging practices that remove a few trees (usually intermediate or suppressed trees whose growth is inhibited by shade from other trees) to make room for neighboring crop trees to fatten. The removed trees are often small and are sold commercially for fence posts, boiler or cogeneration chips, paper pulp, or firewood, or just left to decompose in a precommercial thinning. Thinning generally does not create enough space for regeneration of new trees. Particularly when combined with other practices, such as pruning lower branches on crop trees, thinning is called timber stand improvement (tsi).

▶ **Uneven-aged logging practices** include single-tree selection and group selection. In single-tree selection, loggers remove individual trees and leave the remainder. In group selection, patches of trees up to ¼ acre in size are cut and neighboring patches are left standing. The remaining forest has multiple layers and at least three different age classes. Often the worst-quality trees are taken and the others are left to grow for the future, in much the way disease or insect infestations would naturally thin forests. In order to make a profit from such a timber sale you could choose to also take a percentage of the higher quality trees. A perverse form of selection forestry, which this book discourages, is called high-grading or commercial clearcutting. High-grading removes only the strongest, straightest, genetically superior individuals—the "candy" or "pumpkin" trees—and leaves nearly all the scrappy trees. Another technique, called diameter-limit cutting, also typically destroys the overstory of strong, older trees by cutting everything over a certain diameter. This practice opens up the canopy for regeneration, provides high economic returns, and takes little skill, but it leaves a forest bankrupt of its biggest and best trees.

Ponderosa Pine
Pinus ponderosa

Long evergreen needles, bundled in groups of three. Bark rough and black on young trees, turning platy and yellow brown on older trees.

These are the cowboy trees of the West—they're golden and thick trunked, and they stand a solitary distance from each other. The ponderosa's name derives from the Latin *ponderosus* which means "heavy, weighty, significant." It has a thick trunk and a few top branches that look like flexing arms. Though some ponderosa grow in high, wet areas with grand fir and Douglas-fir, ponderosa pine typically grow in semiarid places that receive less than 20 inches of precipitation annually. The trees develop turnip-shaped roots that reach 2 to 3 feet deep to tap soil moisture. This is in contrast to most other trees that also reach 100 feet or more, which generally have roots only a foot or so deep into the nutrient-rich topsoil. Ponderosa pine often grow with other species adapted to dry areas, such as limber pine, Douglas-fir, western larch, and pinyon-juniper.

The young ponderosa, also called "bull" or "jack" pine, has flat black bark. It matures over one hundred or more years into a "yellow belly," the twisty sapwood drying into valuable heartwood and the bark furrowing and coloring reddish yellow. Wood from mature ponderosa has the same strength and light weight as that of western white pine, making it one of the most valuable softwoods, particularly for window and door frames. In 1999, ponderosa sawlogs typically sold for $500 per mbf delivered at the mill, while other conifers ran around $400 per mbf.

Besides producing high-quality wood, these older trees have thick bark that insulates them against surface fires, which historically swept through ponderosa stands every ten to twenty-five years. Ponderosa saplings are relatively shade intolerant, having adapted to dry pastures where low-intensity ground fires thinned out the competing vegetation. The shade-intolerant ponderosa seeds, like those of western larch, need openings of 10 to 30 feet between trees to regenerate. Fire suppression over the past fifty years and high-grade logging that harvested some of the largest and best pines have allowed shade-tolerant Douglas-fir to germinate in place of ponderosa. Ponderosa are susceptible to both mountain pine bark beetles and the red-needle fungus *Elytroderma deformans*.

In Montana Bill Potter grows ponderosa pines as large as any in the country. "Old pictures show people driving Model-Ts between the trees in this forest," Potter said. "And I want to keep things that way, the way they used to be, without messing things up." During the winter, the octogenarian hobbles around the place with a Stihl chain saw, thinning out the small trees. He only cuts down big yellow bellies when they start to die, even though they sell for twice the price of smaller bull pine per volume. He also leaves some of the dying big trees for woodpeckers, which control bark beetles. Potter isn't getting rich from his logging—he still drives around in a 1970s-era pickup and lives in a modest house—but the logging pays the property taxes on the ranch and puts some money in his pocket.

He also profits from his aesthetic management through the guest ranch on his property. "I'm lucky to have these large trees," he said. "So I only cut them down if I have to. Hell, if I cut 'em, I might never get 'em back." Few places still have trees like this—2 feet dbh, 150 feet tall, 250 years old, yellow barked as if plated in gold—so people name hotels, ranches, and memorial highways "ponderosa," hoping to evoke the lost grandeur of ancient forests.

1

SUSTAINABLE FORESTRY

*"We have more wood in our forest now
than before we started logging."*
— Leo Goebel, Joseph, Oregon

Bob Jackson and Leo Goebel in their Oregon forest

GOEBEL-JACKSON TREE FARM
Joseph, Oregon

ACREAGE
160 acres in northeastern Oregon; NW 1/4 Section 21, T2S, R44E; Leo Goebel and Bob Jackson purchased jointly in 1970; now owned entirely by Goebel

MANAGEMENT OBJECTIVE
"To provide a continuous volume of wood while maintaining a fully stocked, uneven-aged forest."

GEOGRAPHY
5,000 feet elevation, east-facing aspect, loamy-clay soils mixed with gravel

CLIMATE
75-day frost-free growing season
26 inches annual precipitation

STANDING TIMBER VOLUME
Standing volume in 1970: 1,900 mbf (12 mbf per acre)
Standing volume in 1998: 2,200 mbf (14 mbf per acre)

STANDING TIMBER INVENTORY
Douglas-fir: 44 percent of total stems
Grand fir: 31 percent
Ponderosa pine: 15 percent
Western larch: 9 percent
Others: 1 percent
Average tree age: 80 to 100 years
Average tree size: 10 to 12 inches dbh
Average growth rate: 1/5th to 1/10th inch per year
Number of canopy stories: 3 to 4

UNDERSTORY VEGETATION DIVERSITY
Primarily pinegrass-duff with some twin flowers,
orchids, roses, and lady's slippers

WILDLIFE DIVERSITY

Mammal sightings: black bear, bobcat, cougar, white-tailed deer, mule deer, elk, chipmunk, ground squirrel, flying squirrel, snowshoe hare, mice, and vole

Bird sightings: black-chinned hummingbird, black-headed grosbeak, brown-headed cowbird, chipping sparrow, dark-eyed junco, dusky flycatcher, gray jay, mountain bluebird, pileated woodpecker, red-tailed hawk, ruffed grouse, warbling vireo, western tanager, and others

HARVESTING HISTORY

1920s: High-grade harvests of ponderosa pine
1970 to present: Commercial single-tree selection of all 160 acres with variable volumes harvested per year depending on timber prices; 72,000 bf per year average or 450 bf per acre per year (3 percent of standing volume and average growth rate)

ECONOMIC RETURNS

Average gross income from annual harvesting: $30,600 per year total; $191 per acre per year for 160 acres; $425 per mbf per year. Taxes and logging costs are estimated as one-third of gross income.

GOEBEL'S FORESTRY

Leo Goebel owns a mixed-conifer forest in the Wallowa Mountains, just east of the Blue Mountains and the towns of La Grande and Pendleton. He and his business partner, Bob Jackson, have harvested an average of 70,000 board feet (70 mbf) of timber from their 160 acres every year for more than twenty-eight years. They have removed 1.8 million board feet (1.8 mmbf) in total—about 12,000 to 15,000 trees. By choosing to harvest primarily slow-growing and diseased trees, they have given the healthy trees more light and room to grow, and have actually increased the volume of standing timber on their property from the original 1.9 million board feet (1.9 mmbf) to more than 2 million. Goebel harvests less volume than the annual growth rate and chooses trees that are:

- "over-aged" (visibly losing needles and growing less than his average growth rate of 1/8th in./yr.)

- seriously infested with spruce budworm, pine shoot borer, or conk fungus

- genetically inferior with crooks or knotty tops

- crowding taller, higher-quality neighbors

I TRAVELED TO EASTERN OREGON and spent two days with Leo Goebel in Oregon to learn his tricks, to learn how he has more wood now than he had before harvesting. My first night at the Goebels' house, Leo, his wife, Marilyn, and I shared a hearty meal of fried chicken, beans in tomato sauce, mashed potatoes, and dinner rolls. The dining room felt homey. Photos of their five children decorated one wall, and signs with sayings such as "A man who cuts his own wood warms himself twice" hung above the linoleum counter. A television and computer ran in separate corners of the nearby living room. One wall was lacquered with awards Goebel has received for his forestry, including 1992 Western Regional Tree Farmer of the Year from the American Tree Farm System.

After we finished dinner, I went out back to spend the night in a small, green house, which one of Leo's younger daughters had recently moved out of. The house was small and comfortable—the kitchen shelves stocked with boxes of pasta and cans of soup and a dining room table piled with *Mademoiselle, Cosmopolitan,* and *People* magazines.

Before I went to bed, I read a brochure Jackson and Goebel had written about their land. "We feel that a fully stocked, uneven-aged forest with all sizes of trees at all times provides the best utilization of nutrients and water, and best maintains habitat for wildlife and recreation," the brochure read. "Our forest management philosophy is based on the premise that a healthy forest is a complete forest." Instead of harvesting all the plump trees, Goebel cuts a mixture of tree sizes and species so the forest is less vulnerable to species-specific disease and insect infestations.

The next day Goebel and I drove up a short dirt road to his forest. It lies between lowland farms so plowed, fertilized, and sprayed that they look like lawns, and the saber-toothed Wallowa Mountains, as rough as any place in the state. In the forests of the Wallowa foothills, Goebel finds the middle ground between these managed and unmanaged places. He cuts timber while retaining the forest's wild nature—its multiple species and complex architecture.

"I've been in Oregon most of my life," he told me on the drive. "I grew up on a ranch near Wallowa. I graduated from Wallowa High School in the spring of 1949, and one week later my dad loaded all six of us into a 1947 one-ton Chevy truck with a homemade canopy. We drove up the Alcan Highway, still gravel at that time, all the way to Alaska. Dad built the canopy with a bench seat for us to sit on and a window in front so we could see way out over the truck's cab.

"My two brothers went to work for the Alaska Road Department. When the rest of us got to Valdez, we looked for work, too. We walked into a salmon cannery one day, and they put my dad to work on the boiler cooking, my mother to work on the tables cleaning the fish, and my sister to work baby-sitting the younger children. I worked at canning. After that summer, I ended up working three summers in Alaska—everything from loading trucks and doing carpentry to stocking shelves in a grocery store. I later went to college on the GI Bill as a Korean War vet." After college, Goebel said, he taught high school in Oregon for seventeen years. He taught algebra, geometry, earth science, and chemistry.

Two years before Goebel graduated from high school and went to Alaska, Bob Jackson graduated from Iowa State University with a degree in forestry. Jackson worked for the Forest Service for three years, traveling throughout the West. Then he went to work as a timber cruiser, gathering data for harvest planning in the Wallowa Mountains for the same company for whom Goebel's dad worked. Leo Goebel and Bob Jackson met through Goebel's father, and two decades after meeting they bought this land together in the Wallowa Mountains, near where they had first met.

Goebel now owns all the land, but keeps the name Goebel-Jackson Tree Farm. Jackson still comes up nearly every day to thin, prune, pull knapweed, and cut firewood. Sometimes he just walks through the forest, peacefully, as if in prayer.

When we reached the gate to his property, Goebel shoved his blue Chevy truck into park, leaned back in his seat, and started talking forestry. He pulled out articles from the backseat to provide evidence for the points he was making, until the cab became a flurry of papers and Goebel's quick talk.

"I subscribe to about every forestry organization I can," Goebel said. "And I look carefully at what they send me in the mail. Some of these publications emphasize fertilizing the soil to increase productivity. But it's expensive to fertilize an entire forest. Also, I think that the entire soil system needs to be developed, and you do that by building up organic matter, not just adding nutrients. For the most part, you'll be in the square if you just keep organic matter from the trees or crops going back into the soil." Goebel delimbs his trees in the stand before skidding. He scatters the debris, particularly the nutrient-rich needles and small branches, onto the forest floor where it decomposes and enriches the soil. He also leaves medium-sized limbs for squirrels and chipmunks to cache seed in. He collects some large branches in slash piles and burns them to reduce the fire hazard. Trees that fall naturally are left on the ground to donate nutrients to the soil and provide habitat for small animals, which in turn provide more fertilizer in the form of manure.

Goebel's timber harvesting doesn't seriously reduce soil nutrition, he explained, because he only removes 3 percent of the standing tree volume every year, and more than half of a tree's nutrients are isolated in the needles and fine branches that he leaves on the soil. Also, Goebel has not yet seen any live

Douglas-fir on the Goebel-Jackson Tree Farm —Leo Goebel photo

tree needles discolored sickly yellow or green—signs of soil deficient in the primary tree nutrients of nitrogen, potassium, phosphorus, sulfur, calcium, and magnesium. "If I keep the soil partially covered with small branches and needles," Goebel asserted, "then I will be adding organic matter to the soil and will also be keeping the soil surface cool and moist enough to grow more trees."

Goebel keeps his trees closely spaced—5 to 10 feet apart, on average—to protect the seedlings from sunburn and to kill the lower limbs on the larger trees so they'll produce clear timber. "I leave the trees spaced so the boughs just touch," he said. Goebel has mainly shade-tolerant firs, which regrow under the canopy, so he can afford to keep his trees close together without losing new seedlings and saplings. Each landowner, Goebel said, will have to judge for himself the trade-offs between spacing trees close together or wide apart. Close spacing provides a closed canopy for many rare animal and plant species and improves timber quality by reducing branch growth. Spacing the trees wider apart and opening the canopy to sunlight concentrates growth on a few highly valuable individuals, boosts tree vigor to reduce insect and disease infestations, and increases space for regeneration.

Goebel drove me up to a shed where we got on a Honda four-wheeler. I sat on the back, facing backwards, holding onto two rails, with the exhaust putting against the back of my leg. He drove slowly most of the time, but on logging roads he opened up the throttle, and my body bounced so often it felt like one constant motion—bounce-bounce-bounce-bounce-bounce. We leaned into the hills, and he told me to leap off if the machine started to roll.

When Goebel stopped the four-wheeler at an open area, I asked him to tell me, from the beginning, about the evolution of his forest management. "When we first bought this land," he began, "Bob and I did a cruise to see what we really had." The timber cruise, or inventory, started with a series of plots established systematically in the forest, covering about 5 percent of the total property. On these plots, Goebel and Jackson described tree, plant, and animal species, as well as timber volume and density, and then extrapolated to estimate the values for the entire property. Although you should do your own inventory to better understand your property, as Goebel and Jackson did, an inventory is a difficult process. Even a trained forester could spend two full days on a 20-acre inventory. You can get assistance for doing an inventory from state or private consulting foresters (see contacts listed in Appendix I).

When Goebel and Jackson finished their inventory, they found that their forest consisted primarily of Douglas-fir and grand fir, with some ponderosa pine and western larch. A few deciduous trees, such as aspen, alder, and cottonwood, blended with the conifers, particularly in the wet swales. Wheat grasses covered the forest floor, punctuated by calypso orchids, wild roses, and mountain lady's slippers.

Goebel and I stood under the pointed crown of a grand fir, its branches shingling over each other. A couple of crickets chirped in the crunchy brown grass around the tree. A meadowlark whistled farther down the dirt road. From here, the forest seemed incomprehensible—acres of green trees of various species and sizes, with some mix of understory plants. "How did you start to understand what to do with your 160 acres of forestland?" I asked him.

Goebel explained that after he and Jackson inventoried the forest, they developed management objectives. "It wasn't until we really knew what we had," Goebel said, "that we figured out what we wanted."

I later talked to Oregon forest ecologist Chris Maser to get more advice for landowners who want to develop management objectives. "So let's say you decide to remove timber from your land for a house you wish to build," Maser began. "The next step is to figure out how much wood to remove. You need to take a few alternatives, then you need to play each one out. Let's come up with some options for 20 acres: removing 100,000 board feet, 10,000 board feet, and 0 board feet. Think through how each will affect the wildlife, the other trees, the understory vegetation, the finances for your family, the values you espouse in the community, the future choices for yourself and your children and grandchildren. The important part is to be conscious of your actions in the forest—what you do and what the consequences might be."

TECHNICAL NOTE

Forest Types, Forest Succession, and Forest Management

What Kind of Forest Do I Have?

Goebel's Douglas-fir and grand fir forest is a mid-wet forest—one of four primary forest types in the Intermountain region. In the West, forest types are categorized by elevation and moisture gradients. Trees also vary in susceptibility to disturbances and in shade tolerance. For example, mature ponderosa pine have a thick, platy bark that makes them less susceptible to ground fires than Douglas-fir. Grand fir, a shade-tolerant ("st") species, often grows under shade-intolerant ("si") lodgepole and will eventually overtake the canopy. Particularly wet areas may have Douglas-fir growing with hardwoods such as oak or maple. Particularly dry areas will have pinyon-juniper, which grows more like a bush than a tree.

LOW-DRY FORESTS
(3,500 to 5,000 feet, 15 to 20 inches precipitation per year)

Principal Trees	Associated Trees	Understory Plants
Ponderosa Pine (si)	Lodgepole Pine	Snowberry
Western Larch (si)	Limber Pine	Pinegrass
Douglas-fir (st)	Juniper	Ninebark

MID-DRY FORESTS
(5,000 to 6,000 feet, 20 to 25 inches precipitation per year)

Principal Trees	Associated Trees	Understory Plants
Lodgepole Pine (si)	Douglas-fir	Huckleberry
	Western Larch	Twinflower
Lodgepole can grow pure, particularly where occasional fires reopen its serotinous cones, but it is often overtaken by shade-tolerant firs and spruce from below	Ponderosa Pine	Pinegrass
	Grand Fir	Beargrass
	Engelmann Spruce	*There are few understory plants under dense lodgepole stands*

MID-WET FORESTS
(5,000 to 6,500 feet, 20 to 30 inches precipitation per year)

Principal Trees	Associated Trees	Understory Plants
Douglas-fir (st)	Ponderosa Pine	Beargrass
Grand Fir (st)	Western Larch	Devil's Club
Engelmann Spruce (st)	Lodgepole Pine	Serviceberry
Western Red Cedar (st)	Subalpine Fir	Lady Fern
Western Hemlock (st)	White (balsam) Fir	
Western White Pine (si)		

Blister rust has killed most White PIne in the region

HIGH-WET FORESTS
(6,500 to 8,000 feet, 30 to 35 inches precipitation per year)

Principal Trees	Associated Trees	Understory Plants
Subalpine Fir (st)	Douglas-fir	Beargrass
Mountain Hemlock (st)	Lodgepole Pine	Huckleberry
Whitebark Pine (si)	Western Larch	
Subalpine Larch (si)		

Trees at this elevation typically grow stunted or multiple-stemmed and are largely unmerchantable

What Is Succession?

In addition to identifying the type of forest you have, you should also determine its successional, or developmental, stage. Ecological succession is the replacement of one species of vegetation by another, the result of increasing competition for space, light, water, and nutrients. By knowing your forest's successional stage, you can understand its ecological attributes. Primary succession describes the stages of the initial establishment of vegetation in an area. Secondary succession describes the reestablishment of plant life after a major disturbance such as clearcut logging or a severe fire. Below are the succession stages of a typical forest.

Primary succession starts when lichen colonize bare rocks; mosses then replace the lichen and break the rock into soil; grasses and flowering plants commonly grow from this primitive soil, first annuals and then perennials; woody shrubs and trees eventually replace some of the grasses and flowering plants.

STAND INITIATION OR OPEN STAGE
(0 TO 50 YEARS)

During this first stage, the forest is open and sunny. Trees are small and growing precociously. Plant and mammal diversity is also high and includes game animals, such as elk and deer, but also pests, such as gophers and cowbirds, and weeds such as hound's tongue and knapweed.

Stand initiation stage

STEM EXCLUSION OR DENSE STAGE
(50 TO 100 YEARS)

As the forest grows older, it crowds with trees. At this stage, little understory vegetation and few seedlings grow, and fewer animals inhabit the forest, so it is typically a good time for harvesting. Some trees, their growth weakened by competition from surrounding trees, fall to insects, diseases, or wind. Other trees die in the fires that commonly occur during this stage.

Stem exclusion stage

UNDERSTORY REINITIATION OR REGROWTH STAGE
(100 TO 150 YEARS)

This is a period of rebirth during which shade-tolerant trees from the understory grow into the canopy, replacing trees that fell during stem exclusion. Plant and animal diversity rebound during this stage.

Understory reinitiation stage

COMPLEX OLD-GROWTH OR MOSAIC STAGE
(150-PLUS YEARS)

The trees from the understory reinitiation stage grow into middle age. The forest becomes a heterogeneous mixture of young, middle-aged, and old trees, with shade-tolerant and shade-intolerant species. The forest now has more biomass, or total vegetative matter, than at any other stage, with plenty of plant-nourishing organic matter and coarse woody debris. Plant and animal diversity rise to match that of the initiating forest, and this forest supports predatory insects and rare plant and animal species that demand shade and multiple tree canopies. The species diversity and soil richness of the old-growth forest enable it to resist and quickly recover from natural disturbances such as insect infestations, diseases, wind, and fire.

Mosaic stage

Severe natural and human disturbances can alter succession:

- Regular, low-intensity fires have historically maintained ponderosa pine and western larch forests in a unique "savanna" stage with large openings between groups of large- and medium-sized trees.

- In pure lodgepole pine stands, large-scale fires every one hundred years or so keep the forests fluctuating between the initiation and exclusion stages.

- Even-aged logging typically restarts forests at the stand initiation stage, while uneven-aged logging creates a forest structure similar to understory reinitiation or old-growth, depending on the intensity of the harvest and the successional stage of the forest.

- A widespread fire or windstorm can set the forest back to the initiation phase, while isolated insect infestations or disease can move the forest from stem exclusion forward to understory reinitiation.

How Do Forest Types and Succession Affect Management?

FOREST TYPES

Promoting Native Species

David and Kathryn Owen (featured in Chapter 3) favor a mix of tree species that had evolved to the area's natural disturbances and Native American-caused fires over the past 15,000 years, rather than the mix that exists today after a century of intermittent logging and fire suppression. The Owens collect seed from tree species on their property, such as ponderosa pine, that were nearly logged out during the early to middle 1900s. To jumpstart the process they send the seeds to a nursery for germination before planting by hand. To identify the indigenous tree species, the Owens referred to research on pollen samples from modern lakebeds and to Forest Service milling records from the early 1900s. The Owens are now rebuilding their forest into a handsome collection of species that have, over millennia, adapted to the area's soils and natural disturbances.

The Owens' approach is not without its challenges, however. The couple must guess the actual species mix because both the research papers and the milling records give only general information for the valley rather than information specific to their property. Also, the milling records overlook unmerchantable tree species, such as patches of black cottonwood that may have been burned or just left in the quest for more valuable conifers. Planting trees after a harvest is expensive, and the Owens must develop creative harvesting practices or reintroduce fire to keep their planted species.

Forest Succession Following Natural Disturbance Cycles

Logger Bob Love (featured in Chapter 2) recognizes that small disturbances, such as wind blowing over a tree, happen every year, and medium-scale disturbances, such as wind knocking down a group of trees, happen every twenty-five to seventy-five years. But disturbances affecting hundreds or thousands of trees, like a major blowdown or a clearcut, happen only every seventy-five to three hundred years. Two exceptions are surficial, or surface, fires that historically burned through ponderosa pine/western larch stands every fifteen to thirty years, and canopy fires that historically burned in pure lodgepole pine stands every century or so and opened the trees' serotinous cones.

Love emulates the natural disturbance pattern by removing individual trees every year and periodically culling groups of trees; he also harvests dying trees, while maintaining some standing snags and coarse woody debris on the forest floor. When Love harvested

Natural disturbance	Severe wind or fire	Patchy beetle or fungal infestation	Surface fire, wind, lightning strike, old age
Result	Stand dies	Group of trees dies	Individual tree dies
Average rate of occurence	Every 75 to 100 years	Every 25 to 75 years	Every 1 to 25 years
Harvests that simulate natural disturbance	Clearcut or seed-tree harvest	Group selection	Single-tree selection
Comparable successional stage	Stand initiation	Understory reinitiation	Old-growth or shifting mosaic

on the Byrd-Rinicks' property, however, he cut more heavily than usual because the Byrd-Rinicks had a seventy-year-old stand of lodgepole pine susceptible to fire, insects, and windthrow. Love set the forest back to an initiation phase in certain areas so the shade-intolerant species could reproduce, while still maintaining some cover for aesthetics and wildlife habitat.

Forest Succession: Preserving Old Growth

Interior old-growth forests lack the lushness of Pacific Northwest forests, where the trees grow thick and tall and fallen trees make a boardwalk through the wet vegetation. But these inland forests still have stately two-hundred-plus-year-old trees, big fallen timber sprouting new growth, and woodpecker-poked snags. Exemplary landowners in this region save whatever old growth they're lucky enough to have.

Forests become organic regulators when they reach the old-growth stage of succession. Old-growth forests slow water runoff, build soil, store carbon, and provide habitat for a greater array of organisms than younger forests do. Deep old-growth canopies keep the soil moist, which promotes the growth of mycorrhizal fungi threads from the trees' root system. These fungal filaments, which grow up to 1 foot each year, help trees absorb water, phosphorous, and nitrogen from the soil, and the fungi, in turn, absorb sugars from the tree. In some places, trees would not survive without mycorrhizal extensions. Other organisms that prefer old-growth habitat include:

- animals that require specialized habitats, such as salamanders, which live under big logs; lynx and marten, which avoid roads; and woodpeckers, which live in snags

- Neotropical migratory birds, including Swainson's thrush, Townsend's warbler, and MacGillivray's warbler, which feed on the larvae of tree-killing insects, such as mountain pine bark beetle, spruce budworm, and tussock moth

- mycorrhizal understory plants and nitrogen-fixing lichen, which enrich the forest soil

Maser encourages all landowners to seek professional advice from extension services, state foresters, landowner organizations, and consulting foresters, but landowners must first decide for themselves what they want from the forest and what they want the professionals to do for them. "The definition of silviculture is 'the manipulation of forests, through planting, thinning, commercial harvesting, and disturbances such as prescribed fire, to achieve desired objectives,'" Maser said. "But you must decide on the objectives before you do the manipulating."

TECHNICAL NOTE

How Do I Write a Management Plan?

► Consider what you have in your forest, what services you want from your forest, and what you will leave the forest in return.

   Goebel found that he had a diversity of tree species, staggered at different heights in his forest, so he decided to maintain this un-even-aged, multispecies forest in return for a continuous timber supply. Goebel's management plan focused on growth rates and timber supply, but yours could focus on deer and elk populations, tree diversity, wildflower populations, or anything else as long as it is quantifiable so you can monitor it and set achievable goals.

► Detail how you will carry out your objectives, considering what the consequences of each alternative will be in the decades ahead.

   Goebel has developed parameters for when and how many trees he will harvest to maintain his timber supply indefinitely without cumulative damage to the land or financial loss to his family.

► Seek help from forestry professionals, but be sure to stick to your own objectives, specific to your goals and the structure and species diversity of your forest.

   Professional consultants have years of experience evaluating forests and can offer options that landowners may not even think of on their own. On the other hand, Goebel's land is near a national forest that used prescribed burning, but his forest, dominated by old Douglas-fir and grand fir, was not as adapted to ground fires as the ponderosa pine and western larch forests where the Forest Service conducted much of the prescribed burning. He has used selective harvesting without fire to maintain the species composition of his forest.

▶ Divide your property into management units and develop objectives for each of those units.

Goebel has a relatively uniform forest. But landowners commonly have large areas of forest with different tree species and/or trees of different ages. David and Kathryn Owen, for example, divided their lodgepole forest into two sections. In one section, which was high-graded forty years ago, they are thinning out some of the lodgepole to allow the understory of Engelmann spruce, subalpine fir, and Douglas-fir to grow into the canopy. In another, unlogged section, the Owens are logging one-quarter-acre openings to bring in income and to accelerate the succession of the decadent one-hundred-year-old lodgepole, many of which have fallen and tangled across each other.

Goebel and I walked over rocks that stud the grassy forest. In the shade, the soil crunched underfoot with frost, but in the openings between trees the soil was sinky moist. We walked over to a freshly shorn trunk and looked at the rings, alternating light and dark, fat and slim.

Goebel explained that he bases his harvesting system on the growth rate of trees within established plots. He measures the growth rate every five years. He has a number, sometimes a name, for each tree in the plot and measures their diameter on a chart. If the diameter of some of the trees stagnates for two years, then the trees are probably crowding one another and some need to be cut.

"My level of harvesting is sustainable biologically, and it's sustainable economically," he said. "Now, if I would have sold my 2 million board feet of timber back in 1970, Bob Jackson and I would have made about $30,000 gross. Today we could make $500,000 gross because we still have mature trees and not just pole timber out there." Of course, if Goebel had invested the $30,000 from a clearcut in 1970 at 4 percent real return above 3 percent inflation, he would have $480,000 today, nearly the current value of his standing trees, plus the value of a thirty-year-old timber stand. Goebel's investment in trees is clearly beat by other investments that give real rates of return above 4 percent. But Goebel's forest brings many other benefits. He receives an annual income reaching $20,000 to supplement retirement income (a windfall timber harvest would

Spruce beetle, *Dendroctonus rufipennis*

■ STEAMBOAT SPRUCE BEETLES

I wondered what landowners should do when forest integrity isn't enough, when black beetles chew through trees' vasculature, when fungus converts iron fiber to a punky pulp, when fire chars trees black, when wind flattens trees against the ground. I traveled to northern Colorado to find out.

The early morning of October 25, 1997, was snowy and cold—around 0 degrees Fahrenheit—and a storm was blowing in. Ray Heid of the Del's Triangle 3 Ranch should have been feeding horses, setting up hunting camps, and fixing a flat he got on the trailer the other day. The fall hunting season for Heid's ranch is as busy as Christmas season is for city merchants. But instead of doing his regular work, Heid was out in the forest trying to rescue his sons from a hunting camp that was now tangled in uprooted trees.

The wind had hit the Steamboat Springs forest like a sucker punch. Trees develop root systems and even some bark structure to resist prevailing winds. Raging at up to 120 miles per hour, the onslaught struck from the east—due opposite the direction of the prevailing wind, and thus due opposite the direction trees were prepared for.

Engelmann spruce as much as 150 feet tall and 30 inches wide tipped in the wet soil as if pushed over by a giant's palm. Some trees snapped in half like a stick across the knees. Treetops laid across each other like tangled hair; circles of roots stood in the air. Overall, four million trees fell across 13,000 acres. From the air, the blowdown looked like an animal's fur matted down.

Three years later, spruce beetle *(Dendroctonus rufipennis)* began attacking much of the blowdown—60 percent of which is spruce, 40 percent of which is subalpine fir. The beetles aggregated and reproduced in the down wood—millions of tiny black bodies hungry for sugary phloem. The beetles are now attacking live spruce north into Wyoming and south into the Vail Valley.

I went out in the forest with Larry Kent, a silviculturist with the Routt National Forest, to see what the Forest Service is doing about the beetle epidemic. "The beetles are a natural process," said Kent, "partly a consequence of the blowdown and partly just what happens to spruce forests as they age."

The Forest Service is protecting high-visibility areas around the Steamboat Ski Area, at campgrounds, and along highways. "We're not trying to stop the epidemic everywhere," Kent said, "but in some places we are trying to suppress the beetle populations so that natural predators, such as woodpeckers and wasps, can control them."

In live spruce stands, the Forest Service plans to thin across all age and size classes, reducing the density by one-third. The increased air movement in thinned stands reduces the concentration of aggregating pheromones, special chemicals that the beetles secrete. These chemicals attract thousands of beetles to gather and attack a tree at once, overwhelming the tree's pitch-out defenses.

In other areas, where logging is not economical, the Forest Service is felling spruce trees as bait for the beetles—about one tree for every ten to twenty infected trees where the beetle populations have not yet built up. After allowing time for the beetles to aggregate in the fallen trees over the spring and summer, crews cut the trees into 3- to 4-foot-long chunks so each side of the tree can be turned over. Then they skin the bark off. The beetles cannot survive outside the blanket of bark, despite their adaptation of excreting water and manufacturing glycol as an antifreeze.

Kent and I drove up Buffalo Pass, a scenic road north of Steamboat. We met Kenny Bigback, part of a temporary crew that had come from the Northern Cheyenne Reservation outside of Billings, Montana, to work on beetle control in a patch of blown-down trees. Bigback worked the bark off a block of spruce, revealing the blonde wood beneath the dark bark. The end of his chain saw had a "log wizard"—a set of planer blades that gnaws the bark from trees as the saw moves up and down the tree.

Bigback's wife, back on the reservation, had just given birth to a baby boy. Bigback said that he misses his family sometimes, as he moves throughout the West on Forest Service contracts, but he has chosen this work despite having a college degree in education. "It feels good to work hard," he said. "I like walking back to the trucks with everyone at the end of the day when we're all so tired that no one can talk."

On the way back to the road, Kent and I talked about what he might do if he were a landowner faced with a spruce beetle attack. Depending on his objectives, he might just let the infestation run. Or he might protect his timber and reduce the density of the forest by harvesting susceptible spruce. "If I owned forestland, I could spend a lot of money trying to keep things the way they are," he said. "But forests are dynamic, always changing, so it's important to identify the changes taking place and work with them."

have involved higher taxes). He can also cut heavier during years of high timber prices and cut lighter during years of low prices—much like a farmer who
stores grain to maximize profits. Overall, Goebel is "working on land value,"
he says. He can enjoy the view and summer walks, and his land will be a
handsome inheritance for his heirs. According to western real estate agents,
fully forested land sells for three to ten times as much per acre as comparable
clearcut land (see Chapter 3 for more information about estate taxes and the
economics of sustainable forestry).

Goebel first removes trees that are dead, diseased, deformed, damaged, or
shaded out of the canopy—trees likely to die anyway. Then, if he still hasn't
filled his volume quota, he removes healthier trees that may eventually crowd
each other out—"a little steak to add to the hamburger," as he puts it, or valuable trees to pay for the cost of harvesting low-value trees. Most of the trees
Goebel cuts are around sixty to eighty years old and 10 to 14 inches dbh. These
middle-aged trees contain merchantable quantities of wood, unlike the young-
est trees, and they are still relatively plentiful, unlike the oldest trees. Goebel

TECHNICAL NOTE

Forest Health

What Has Happened to the Health of Western Forests?

Intermountain forests, particularly dry, low-elevation forests of ponderosa pine, western larch, and lodgepole pine, regularly burned in fires
that lightning or Native Americans sparked over thousands of years.
During the past fifty to one hundred years, those fires have been
snuffed, allowing trees to grow into dense matchbook forests susceptible not only to fire but also to windthrow, insects, and disease.

Some believe in prophylactic harvesting—that is, harvesting trees
aggressively to try to prevent fires or insect infestations. Others think
forests should just be allowed to face natural fires and disease infestations that clean out weak trees and built-up wood. The best solution
is a middle course of restoration thinning—cutting only the small or
weak trees that fire or disease would likely take. The following series
of diagrams show how western forests have changed over the past
two centuries.

Raging flames from fires in 2000 in the Bitterroot Valley of Western Montana. After decades of fire exclusion, forest fires have become so intense in recent years that they have burned 200-year-old trees that had survived generations of low-intensity fires.

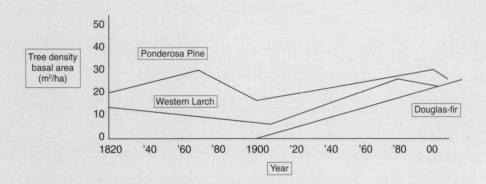

Overlogging of ponderosa pine and western larch, along with overgrazing of their young seedlings, have combined with fire exclusion to allow the eastern Cascade Mountains of Oregon to fill in with shade-tolerant, fire-intolerant species such as Douglas-fir and grand fir. Not only are Douglas-fir and grand fir more susceptible to insect and disease infestations than ponderosa pine and western larch, but the numerous tree species now crowd the forest so all trees become more susceptible to health problems. —Data from Agee, 1994. Graph based on computer simulations that connect intermittent field data from Oregon

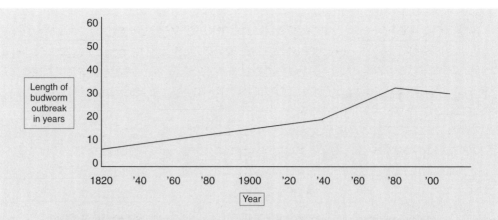

As Douglas-fir and grand fir overtook low-elevation ponderosa pine–western larch forests, spruce budworm infestations became more intense and longer-lasting. —Data from Anderson, Carlson, and Wakimoto, 1987. Graph based on data from western Montana

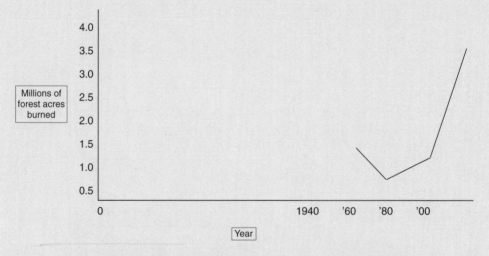

Large-scale fires then swept through forests crowded with fir and defoliated by spruce budworm. —Data from Taylor, Oliver, et al., 1997. Graph based on field data from all of western U.S

What Should I Do to Keep My Forest Healthy?

A modest forest density will guard against severe fire and insect in-festations and protect your investment in timber. Selective harvesting can provide more growing room for healthy trees, allowing them to maintain their strength, and it reduces the number of trees suscep-tible to or already infected with insects or disease, without com-pletely sanitizing the forest. Where mountain pine beetle, tussock moth, dwarf mistletoe, and other pests have a strong foothold—

afflicting well over 1 percent of the standing volume or 25 percent of the growing stock—aggressive group selection or even-aged harvests may be needed to cleanse the forest and develop a vigorous new stand of trees. Of course, the forest can always be left to die and regrow at its own pace.

If you decide to selectively harvest dead and dying trees to fortify your forest against future infestations or fire, you must ask:

WHAT IS AFFLICTING THESE TREES?

Symptoms	Insect/Disease	Tree Species Affected
Red needles and red-brown boring dust at base of trees	Bark beetles	Lodgepole Pine Ponderosa Pine Douglas-fir Engelmann Spruce
General low vigor and rings of dying trees (bull's-eyes)	Root rot	Douglas-fir Subalpine Fir Grand Fir
Deformed branches with curled ends like witches' brooms	Dwarf mistletoe	Lodgepole Pine Western Larch Douglas-fir (occasionally)
Conk fungal outgrowths	Stem rot	Grand Fir Western Hemlock Subalpine Fir Douglas-fir Engelmann Spruce (occasionally) Western Larch Ponderosa Pine
Webbed silken nests on branches	Western spruce budworm	Douglas-fir Grand Fir Subalpine Fir Engelmann Spruce (occasionally) Western Larch
Frothy cocoons around needles	Douglas-fir tussock moth	Douglas-fir Grand Fir Subalpine Fir Engelmann Spruce
Shortened branches and needles on tops and ends of branches	Western pine shoot borer	Ponderosa Pine Lodgepole Pine

Are These Trees Really Dying?

Deformities, bark blisters, red needles, pitch tubes, and twisted branches all flag potential diseases or insect infestations, but you should examine the individual trees to see whether they are actually losing growth. Some resistant white pine, for example, express blister rust but continue growing strong. If these trees are saved, they can pass on resistance to new trees that grow from their seeds. But if these visibly infected trees are growing poorly and interfering with the growth of merchantable trees nearby, they should be harvested. One rule-of-thumb for vigor holds that if the live crown ratio—the percentage of total height of a tree with green branches—is less than 33 percent (in other words, less than a third of the tree has green branches), then the tree is dying.

Fungal conk on a grand fir tree in Montana

Western larch snag on J. R. Reynolds's property in Montana; a woodpecker roost carved in the top is now used by an owl.

Which of These Dying Trees Should Be Harvested?

Some dead and dying trees need to be saved for wildlife. Woodpeckers living in snags can also check small insect outbreaks. Forest Service researchers recommend that you leave one snag greater than 12 inches in diameter and greater than 15 feet tall every two acres to maintain near-maximum populations of bird species—such as the common flicker, northern three-toed woodpecker, and hairy woodpecker—that need hard snags for nesting. The most important tree species for wildlife snags are ponderosa pine, western larch, cottonwood, and aspen.

In addition to preserving dead standing trees, Forest Service researchers suggest you leave at least two dead fallen trees greater than 12 inches in diameter and 4 feet long on each acre. Fallen wood or coarse woody debris can provide a substrate for the growth of new trees; cover for insects, amphibians, reptiles, and small animals; and a filter to slow runoff in the spring.

How Should the Dying Trees Be Harvested?

Plan your selective harvest so that it:

- retains large, healthy trees
- reduces the population of infected trees and susceptible tree species
- maintains a diversity of tree ages and species to make the forest less vulnerable to future attacks
- saves standing and down deadwood

However, selective harvesting can aggravate some forest health problems by:

- scraping trees and compacting the soil during harvesting, which can expose trees such as grand fir and Douglas-fir to fungal infestation and rot
- leaving too many small branches and needles on the forest floor, which can increase fire hazard and attract beetles
- retaining multiple canopies where insects such as spruce budworm and parasitic plants such as dwarf mistletoe can drop into young trees

Know the extent and nature of health problems in your forest before deciding on a harvesting system, if you choose to harvest at all.

always keeps some trees more than 150 years old, cognizant that the economic value of these trees improves the property's value. The trees also perform ancillary ecological functions, such as providing shade to young saplings, dispersing hardy seed, giving refuge to hiding animals, and, later when the trees die, donating habitat to woodpeckers and nutrients to the soil.

Goebel and I got back on the four-wheeler and bounced down the road into an open, grassy area—a space in the shade-dark forest. "This is old Riley's place," he said. "He had a moonshiner's cabin in this opening in the 1920s." He shut off the engine. We sat in the quiet for a while. I thought about Goebel's harvesting methods and carefully worded my next question.

"Now you told me that you save the oldest and straightest trees and cut those that are crooked or forked or somehow screwed up," I said. "But shouldn't good forestry involve good husbandry, looking especially after the bent and misshapen children, so forests don't become artificial, just plantations of straight trees?"

Goebel thought a minute. "You're right, but I don't manage it like a plantation," he said. "There are a lot of places where I leave a misshapen tree for a bird snag and other places where I decide to remove one of the larger and straighter trees to open the canopy for new tree growth or to finish off an otherwise unprofitable thinning operation."

Goebel pointed to a fallen old tree, circled by new saplings. "Leavin' that tree lay there isn't a waste," Goebel said. "It's an investment, bringing up those young saplings with nutrients and moisture."

Near the fallen Douglas-fir stood alder. Some landowners in this region have cut or sprayed herbicides to eliminate deciduous alder and make room for more valuable Douglas-fir. But the alder's roots contain a nitrogen-fixing bacteria that converts atmospheric nitrogen to usable nitrate. The nitrate leaks out of the root nodules to enrich surrounding soil. Alder also reduces the spread of *Armillaria* and *Phellinus* fungi by breaking up the root systems of neighboring Douglas-firs. Finally, hardwoods such as alder have few volatile oils in their leaves, so they are more fire-resistant than softwoods. "My alder shows that you shouldn't manage a forest just for efficiency," Goebel said. "You should also manage it for integrity."

One morning, Goebel and I ate at the Cheyenne Cafe, a small place in downtown Joseph with brands from local ranches burned into the wooden walls. People meet here with such regularity that everyone looked up at me when I came in.

Goebel and I sat down with two of his friends. Across from us sat Squirrel — a thin, young logger wearing a plaid shirt, suspenders, and an angular goatee. He makes his living climbing up trees to prune their limbs, or cutting off big tops so the trunks don't break when they're felled. "Squirrel can climb trees faster than anyone in the county," Goebel said proudly. "That's how he got his nickname." Goebel's other friend, Russ, was a beefier man who wore a straw cowboy hat, a T-shirt, Wrangler jeans, and Red Wing work boots. Goebel told me Russ cuts trees so fast that he tapes the throttle of his Husqvarna open, so after felling he can run up and delimb and buck the tree without turning off his saw. Russ wears through two saws a year.

That day was hazy for the third straight day in rural Oregon. The newspaper blamed an inversion weather pattern and high moisture, but everyone at the table was certain the smoke came from prescribed burning. Managers of both private and public forests in the region increasingly use prescribed burning to reduce fire hazards from fuel loading—the buildup of dead wood on the forest floor and the dense stocking of small trees—and to restore the ecological character of the ponderosa pine forests.

"I can't believe that there are people around here going around and burning valuable timber," Goebel said angrily. "And they're causing air pollution!"

"Yeah, I'd pay them for that timber," Russ said. The fires are typically surficial, I said to myself, and burn young trees that are too small to harvest economically.

"All that slash and dead material they burned would have done some good for the forest," Squirrel said. "It's important to have that material to retain water and nutrients in soil." When applied properly, prescribed fire is low-intensity, I thought, and it will not burn large logs. Fire can actually accelerate the incorporation of small pieces of wood into the soil.

The café was small and unfamiliar, so I stayed quiet with my thoughts—Goebel and I would talk later. I just ordered hash browns and dollar cakes. The pancakes came quickly and were so huge the edges spilled over the side of my plate. "I can tell you haven't eaten at the Cheyenne before," Squirrel joked. "Whenever I eat here I order the children's portions." When I later paid the waiter, he told me the Cheyenne took pride in giving its customers their money's worth.

Goebel and I said good-bye to his friends and walked outside to Goebel's truck. "Prescribed fire's hard to apply," I said, working up to my thoughts. "It has plenty of drawbacks because it's easier to get wrong than right. Still, burning can be good. It can stimulate germination of shrubs like willow for elk to feed on, reduce hazard fuels on the ground, and provide open places for shade-intolerant trees like ponderosa pine to grow."

"Yeah?" Goebel asked, cocking his jaw toward me in a challenge. "I'd like to see a good prescribed fire."

The Arnos, Fire Brothers

The photo I took of Nathan Arno outside of Stevensville, Montana, could be the photo of a college football player. He's smiling and young, with brown hair and a face that shows both self-confidence and friendliness. But he's out in the forest instead of a stadium, and he's holding a drip torch instead of a football.

Arno founded Wildland/Residential Fire Management, a consulting group that harvests small trees to reduce fuel loading, then sets prescribed fire to clear up the dead material on the ground. "We usually work out a trade with the landowners," Arno told me. "We harvest merchantable trees, then use the proceeds to fund our work in another area." Some landowners get funding to support thinning and prescribed burning by applying for federal Stewardship Incentive Payments (SIP), which provide assistance to landowners for otherwise unprofitable forest improvements. (Information about the Forest Land Enhancement Program, which recently replaced SIP, is available from state foresters' offices listed in Appendix I.)

So here's the deal: you carry homeowners insurance that covers fire losses and sign a contract waiving Arno of any liabilities. Then Arno comes out in the spring or fall with his chainsaw, a farm tractor for skidding logs, a four-wheeler with a fireline plow, a camper trailer, a pickup with a water tank and pump, and aluminum drip torches. "Timing is usually the hardest part, and I won't burn unless conditions are just right," Arno said. "But when it's dry and cool, I'll burn up to 40 acres a day. I usually camp on the property the first night and get up regularly to check that the fire hasn't reignited."

Arno told me his steps for setting prescribed fire. After choosing a time and an area for burning, he moves big piles of litter away from larger trees and cuts down smaller trees (less than 6 inches dbh) that might send fire up into the canopy like a ladder. "Next I establish firelines, at least 18 inches wide and down to mineral soil," he said. "I have one or two people monitor the lines, then I set the fire in strips, burning into the wind." He suggests landowners keep their lawns green and watered around their buildings as an additional precaution. He is cautious about burning on dry southern and southwestern aspects and tends to burn on fairly level ground. Arno pays attention to changes in wind speed and direction and always has an emergency plan.

"Some say that the fire just kills trees and adds to air pollution," Arno said. "But you're going to have air pollution one way or another, either with low-intensity fires like this or big wildfires. Low-intensity burning has been nature's way of taking care of things. Now, we're trying to emulate nature without destroying houses." For safety, Arno burns opposite to nature, during cool, calm periods, particularly in the fall when the burns are less likely to damage nesting birds or germinating plants.

Arno gave me a can of diesel and gas, and I spilled it on the ground, then lighted it and watched the fire expand—whoomph—and collapse back in. I warmed the wick from an aluminum drip torch along the fire and ran the torch back and forth in strips from uphill to downhill. The uphill fire would burn into a trench, and the other fires would snuff into the char of already burned strips.

On the ground, the fire slithered and slunk through low branches and over fallen wood. Sometimes it would leap up maybe a couple of feet, but mainly it moved along the ground like an eraser, turning the green and brown forest floor to blackboard, scuffing the bases of trees. This wasn't a wildfire, some

Prescribed fire in a ponderosa pine forest —Photos by Scott Chase, Vermejo Ranch, New Mexico

hungry animal lunging through the woods. This fire was tame. After fifteen minutes of burning, I threw dirt on the flames and scuffed them out with my boot.

Nathan's older brother, Matt, is also a consulting forester, and the two grew up learning how to fell trees and light prescribed fires just as other country kids learn how to milk cows and stack hay. Their father, Steve, a retired Forest Service ecologist, has used thinning and prescribed fire for over two decades on his ponderosa pine property. Both sons have business territory around Missoula, where they went to forestry school, and there's a cardboard mill that buys small wood for pulp. Nathan has the Bitterroot Valley south of town and Matt has the Blackfoot Valley east of town.

"I look at each property as an ecologist, paying attention to all the resources from wildlife to vegetation to soils to water to trees," Matt Arno said, describing his work. "Everyone working with me from my partner to my subcontractors looks at the forest holistically." Arno works on a bigger scale than Nathan, typically working on ranches rather than homesites. He does the forestry himself, determining which trees to take and marking them with tape; he then hires out the logging labor. He interviews subcontracting loggers carefully, and sets specific guidelines for leave trees — diameter, crown size, canopy position, vigor, species diversity, and grouping. He also has specific penalties for damaging the leave trees, allowing only one small scar per acre and one big scar per 10 acres.

"The premise of ecologically sound forestry is to leave the good trees," he said. "The question is which trees are the good ones." The number of good trees he leaves depends partly on his judgment and partly on the forest he has to work with. Arno and his partner come into a forest and mark the big crop trees that they want saved, the trees that would have survived a burn. "We stand back and look at the trees and treetops and figure out what the forest would look like if it had regular surface burns," Arno said. "We often ask each other for second opinions, ask whether this really looks right from old photos we've seen and other forests we've looked at."

In tree thickets, Arno has cost landowners up to $500 per acre, while in forests stocked with big ponderosa pine he has made them up to $500 per acre. He has worked on tracts ranging from 1 acre to 400 acres. "I can tell in the first conversation whether I want to work with a landowner," he said. "Do the landowners just need some cash or do they have a vision for what they want their land like, do they want their forest shaped up?"

Arno's clients have low-elevation ponderosa pine, western larch, and Douglas-fir forests. Unlike tropical forests, which are saved by leaving them alone, these forests are saved by going in and doing something. These forests, off their natural burn cycles, have grown into rats' nests. Arno restores the architecture of these forests so they can again burn with regular ground fires without burning down.

Arno took me to a forest near Philipsburg, southeast of Missoula. The round hills here are blonde in the fall, so other colors come out strong: the blue of the water looking turquoise, the green of the trees looking emerald.

Arno pulled his diesel Ford off a highway and moved up to the gate of a ranch. The forests above the pasture and hay fields were pure Douglas-fir, higher than ponderosa pine forests but still adapted to surface fires every decade or two. We got out and hiked up to a ridge to look at an area he had cut and burned. "This is the spindly stuff we had to work with," he said, showing me a drainage clogged with small Douglas-fir. He kicked at the forest floor and uncovered layers of dead wood, branches, and needles. "And this is why we have to burn," he said, "to take off the material built up on the ground that would carry fire like kerosene." Prescribed fire, Arno said, also reinvigorates woody browse like willow and maple for wildlife and opens seed space for new trees and plant species.

The loggers came in and removed about a load per acre here, half of it big enough for sawlogs and the other half pulp. On big projects like this, more than 30 acres, Arno uses a cut-to-length harvester to cut and buck the wood in the forest. He then forwards it in bundles to the landing by the road. Arno prefers the forwarder machine, despite its relatively large size and expense, as it's quick and eliminates the spiderwork trails created by repeatedly skidding cut trees out to landings.

After harvesting, Arno sets "jackpot" burns in scattered slash piles rather than geometric lines of prescribed fire. On small properties and near houses, he chips the slash and dead wood on the ground, rather than burning it. A few trees burn too much from the jackpot fires, making them susceptible to beetle attacks, but most trees don't burn at all and the ground is opened up for grasses, herbs, and other trees such as aspen. The forest ground after a jackpot burn follows the variegated surface burns in nature, Arno believes, where it looks like a reptile partway through the molt.

Arno and I walked back down the ridge, got in the truck, and drove to Missoula, going through successive stages of road from field tracks to dirt roads to paved roads to interstate. Arno pointed to a hill that had been clearcut after a severe fire so only a few burnt trees, charcoal sticks, stuck up from the mountain. He looked at me: "I wish that I could do more for this country."

Before I left Goebel's place, I felled a tree with Goebel's saw and helped him skid it. I've cut fewer than a dozen trees in my life, and I'm still astonished at the feel of the saw in my hands, its heaviness and hunger. The yellow McCullough had a medium-sized bar, 20 inches long. I started it as I would start a lawn mower by opening the throttle, pressing the machine against the ground with my foot, then jerking up on the starter cord. After a few sharp

TECHNICAL NOTE

Fire

Why Should I Burn My Forest?

It's a Montana spring day of jacket mornings and evenings and T-shirt afternoons. Fire ecologist Steve Barrett of Kalispell, Montana, and I are walking together through a mixed western larch/Douglas-fir forest. His words are gunshots. "The infilling here is severe," he says. "This forest is going to hell."

He describes the changing forest. The western larch here used to be 350 years old, on average, but now are only 80; the density of Douglas-fir used to be 10 per acre and now exceeds 300; the total forest density used to be 40 to 50 trees per acre, but now reaches 500 to 600.

"We have dramatically changed our forests with twentieth century fire exclusion," Barrett says. "We've let the forest fill in with little trees that sap water and nutrients from the bigger trees and bring

Photos taken near Sula, Montana in the 1890s (top) *and 1980s* (bottom) *show the effects of nearly a century of fire exclusion.* —Courtesy of U.S. Forest Service

low-intensity fires up into the crown. Fire thinned out species like Douglas-fir in the past, but now these species clump up the forest."

Fire scars on old stumps show that before extensive fire control by trucks and helicopters in the 1900s, dry Rocky Mountain forests had small, surficial fires every 15 to 25 years that killed small trees and groundcover. Researchers speculate that aboriginal peoples who came into the continent after the last glaciation, 10,000 to 15,000 years ago, started many of the fires in the dry, lower-elevation forests. In the early 1900s, for example, Forest Service rangers reported seeing Kootenai Indians regularly burning for game drives, warfare, horse and game habitat, and trail and campsite clearing. Vegetation in the West adapted to regular fires; carbon scars on tree rings dating back to the late 1500s show that, on average, 6 million acres burned each year in the Columbia River Basin, but less than one half million acres have burned each year over the last half century.

"We could just leave things the way they are, with our forests clogged with scraggly trees," Barrett says. "But since we screwed things up from the way they have developed for the past few thousand years, I think we have a responsibility to fix them."

Barrett recommends harvesting at least half of the trees less than 100 years old in a ponderosa pine forest, particularly Douglas-fir, grand fir, and subalpine fir. Nearly all of the older trees should be saved. Then the forest should be burned.

Burning and thinning are not suitable for all forests, however. Prescribed burning can damage shrublands and high elevation forests that had a history of irregular crown fires rather than regular surface fires. Barrett has developed this scheme of historical fire disturbances for different forest types:

FOREST TYPE	HISTORICAL FIRES
Parklike Ponderosa Pine forests	Surface fires every 10 to 25 years
Ponderosa Pine/Douglas-fir/ Western Larch forests	Surface fires, some crown fires every 25 to 40 years
Douglas-fir/Western Larch/Grand Fir/ Engelmann Spruce forests	Crown fires, varying intensity every 70 to 100 years
Lodgepole Pine forests	Intense crown fires every 150 to 300 years
Subalpine Fir/Subalpine Larch/ Whitebark Pine forests	Variable fire intensities every 40 to 300 years

Barrett lists the ecological advantages of using prescribed fire, marking each point with a finger of his hand:

- Prescribed fire reduces the thatch of litter and duff that keeps seeds from germinating. Thatch keeps the soil moist and cool for regeneration of shade-tolerant fir but stifles the regeneration of shade-intolerant ponderosa pine.

- Prescribed fire thins out patches of small, shade-tolerant fir that can't be economically harvested. Ground fires, however, can also kill older trees that have accumulated duff around their trunks unless the duff is raked away.

- Prescribed fire reduces the live and dead fuels that promote catastrophic fires. Ground fires must be kept small and controlled, however, so they do not volatilize soil nitrogen to vapor, and do not burn large woody debris, which provides nutrient-rich nurseries for seedlings.

- Prescribed fire reinvigorates woody browse such as willow and maple for wildlife including rabbits, moose, and elk.

Barrett's dark eyes turned more intense as he finishes explaining himself. "There's simply no way to replicate fire without using fire."

How Can I Protect My Property from Wildfire?

Wind and fire act synergistically with embers flying out to cause new ignitions, hot air drying out vegetation ahead of the blaze, and fire creating its own turbulent air flow. The flame can then sometimes become a "blowup," the rare but dangerous event where flames expand like gigantic arms and enwrap hundreds of acres in minutes. The "Big Blowup" in Montana and Idaho in 1910 killed almost 80 firefighters and spurred the country's fire suppression policies. The 1910 fires were the most deadly and the fires of 2000 were the most severe, burning over 7 million acres across the country.

Intermountain forests are more combustible than ever. You can reduce your susceptibility to wildfires with the following strategies.

▶ Selective Logging

To minimize fire danger, the Forest Service suggests a spacing of 15 to 25 feet between trees or a basal area of 50 to 100 square feet

Fire ecologist Steve Arno, Nathan's father, stands next to willow that has regrown after prescribed fire on his property.

per acre in ponderosa pine forests. You should harvest primarily grand fir, Douglas-fir, subalpine fir, and Engelmann spruce, which were historically intermittent in these lower-elevation forests. Be sure to dispose of the logging slash and the dead wood on the ground by hauling it off for firewood, chipping it, or burning it. Also be sure to harvest small trees, or "ladder trees," approximately 4 to 10 inches dbh, which could run a ground fire up into the canopy. During the 1980s, boom years when loggers removed record numbers and sizes of trees, wildfires burned more acres on average than in the 1970s or 1990s. Aggressive logging can actually boost fire hazard by removing some of the biggest trees resistant to fire, stimulating dense regeneration, leaving flammable piles of slash, and opening up the canopy to drying sun and wind. Most of the trees that need to be removed to reduce fire danger have little commercial value, but clearing them out of your forest will decrease the risk of catastrophic fire and improve the overall health of your forest.

A typical ponderosa pine forest in Montana clogged with tree fuel

The ponderosa pine forest on the E-L Ranch in Montana, where smaller trees have been thinned out and old trees still stand, their thick bark protecting them against fire

▶ Smart Homebuilding and Landscaping

Keep the ideas below in mind when planning and maintaining your home and yard:

House Design
• Avoid building on steep slopes, ridges, or at the ends of narrow canyons—places where fire naturally accelerates.
• Use slate, tile, asphalt, concrete, steel, or tin roofing rather than wood shakes.
• Eliminate overhanging eaves.
• Line the base of the house with rock, brick, metal, or adobe siding.

House Maintenance
• Clear vegetation beneath decks, and clean gutters.
• Store firewood away from the house.

Access
• Build major roads and driveways wide and sturdy enough and with adequate turnarounds for fire engines.

Landscaping
• Keep lawns well watered.
• Landscape with plants that have high moisture content and less volatile oils in their leaves, such as Kentucky bluegrass, blue fescue, crested wheatgrass, western wheatgrass, rye grass, or maple, birch, aspen, cottonwood, or willow.
• Keep trees 10 feet away from the house; prune trees within 100 feet of the house to 15 feet high and spaced 10 feet between crowns.
• Keep ground debris less than 3 inches thick.
• Develop a reliable water source and purchase a generator that can pump the water.

▶ Prescribed Fire

Prescribed fire is probably the most effective management tool for controlling wildfire, particularly in ponderosa pine or western larch forests, which have adapted to frequent, low-intensity burns. However, prescribed fires involve considerable risk. You will need expert help in setting the fires and will have to assume your own liability under your homeowners policy. Nathan Arno warns that you must:

• Remove some duff from around the base of larger trees.

• Harvest "ladder" trees from the forest.

• Apply to the appropriate government agency for a burning permit.

• Set small fire strips that run against the wind.

• Dig trenches down to the mineral soil around the peripheries of the area to be burned.

• Set the fire only on a cool and calm day.

• Start small, only burning 1 to 5 acres.

• Extinguish the fire if the wind speed or direction change dramatically.

• Develop an emergency plan which includes an adequate and accessible water supply.

For more information on protecting your home and property, see *www.firewise.org*, an interagency government website.

coughs, the engine caught. I lifted the saw and gassed it a few times. I glanced over at Goebel, and he nodded back at me. I looked again at the open slot where I wanted the tree to fall. I brought the saw against the tree.

I made a "face cut" first, wearing the tree open with a 45-degree cut and a straight 90 degree cut below, so they made a wedgelike slice as they met. Then I moved to the other side of the tree and aligned the blade above the wedge. I put the saw in for the back cut, just a quick horizontal slice on the side of the tree opposite the face cut, and kept my eyes split between the tree and the saw as I cut. I pulled the saw out and moved away as soon as the tree quivered. The balsam fir leaned forward, hesitated, then collapsed in an explosion of color and noise.

I stood with the saw shut off and dangling from my hand. My booted feet rested on the soil. The sun was still high. There was an empty space where a growing tree had stood.

But I took wood — a material so much cheaper and more alive than metal, so much less polluting and more renewable than plastic. This work was dirty and loud, but it felt right.

"Now let's get that tree out to the deck," Goebel said, jarring me. Goebel walked over to the skidder, a machine used to drag cut trees from the forest to a clearing or landing where it can be picked up by a truck. He started the machine's diesel engine and ripped the crawler-tractor he used as a skidder over the soil to the tree. While Goebel moved the skidder, I started the saw again and clipped limbs off the main trunk, the saw's momentum carrying itself. Some branches fell onto the ground limp, and others shot off fast.

After I had delimbed the tree, Goebel released the winch on the skidder. I pulled out a thick cable and fixed it around the tree. I set a nooselike choker by wrapping the cable around the tree and putting a plug at the end into a clasp that ran freely along the cable. Goebel asked me if I was ready. When I signaled, he turned on the winch, with a pulling power of 32,000 pounds, and the choker tightened, its cable reeled, and the winch pulled in the tree. The metal machinery squealed.

Goebel drove the skidder forward, stopping to get two more trees he had felled earlier. I pulled my cap down and covered my eyes as he axed away branches from one of the trees so he could easily move it. He fixed these two other logs to the cable, winched them in, then got back in the cab and engaged the clutch. The engine pulled, dragging the trees behind. Goebel drove the skidder through a geometry of trees and tree branches back to the landing. The machine moved, sounding cranky, over ridges and slopes, up and down the pitted road.

Finally, Goebel drove up alongside a deck of trees. He loosened the cable so the trees' butt ends, partly suspended when he drove so they wouldn't burrow into the soil, fell down to the ground. I released each log from the choker cables, then Goebel pivoted the machine so he could pile the logs together with the blade. He piled the logs, stopped the machine, hung his metal hat on the stick shift, and got down.

We stood together, smelling of sweat and diesel and sawdust. "Why aren't there more landowners like you who really think about what they're doing when they log?" I asked.

"It's as hard to bank trees for the future," he said, "as it is to save money for the future."

Lodgepole Pine
Pinus contorta

Fairly long evergreen needles, bundled in groups of two.
Bark thin and scaly, colored light brown.

Lodgepole pine characterizes western forests as sagebrush does western prairie. This thin tree, used by Plains Indians for tepee poles, ranks third in total land coverage in the United States behind Douglas-fir and ponderosa pine. You can easily identify lodgepole by its long trunk, delicate, flaky bark, and bundles of two needles. Another version—short, shrubby, and contorted by wind—grows along the Pacific coast.

Lodgepole is the most widely distributed conifer in the West, growing from elevations of only 500 feet to as high as 11,000 feet, in deep loess and in bare gravel, in areas of less than 10 inches of annual precipitation to ones that receive 200 inches. It is the only conifer species native to both Alaska and Mexico. Lodgepole's fecundity, along with its stalwartness, has enabled it to grow in such diverse places.

Many lodgepole bear serotinous cones—cones that contain a resinous glue that keeps them from opening until exposed to temperatures exceeding 100 degrees Fahrenheit. Thousands of cones can build up on the forest floor, then explode in thick growth after a fire, in densities ten times those of other conifers. Lodgepole historically faced catastrophic fires every one hundred years. The famous Yellowstone fires in the summer of 1988 allowed the lodgepole's serotinous seeds to open and sprout a new forest.

Lodgepole are short-lived, rarely lasting more than 150 years. Because they are shade intolerant, many lodgepole die as other trees overtake them from below. Other lodgepole become infected with mountain pine beetles or parasitic mistletoe. Others, thin-stemmed and thin-barked, fail in high wind or in fire or under heavy snow. Lodgepole is sometimes a permanent resident, but more often is a pioneer, coming quickly into a place and building a shady shelter under which fir and spruce grow.

Until the 1970s, dense stands of thin lodgepole were burned just to get at more cherished timber. But now foresters recognize lodgepole's important ecological and economic roles. It grows back quickly after a fire or harvest, protecting the soil, providing cover for small mammals from predatory birds, and building a canopy for shade-tolerant trees to grow under. It grows with minor taper, so even the uppermost parts of the tree can be used as merchantable wood.

ZEN LOGGING

"It's not what you take from the forest
but what you leave that's important."
— BOB LOVE, COLUMBIA FALLS, MONTANA

Bob Love and his dog Murphy in front of a
forest he logged in northwestern Montana

BYRD-RINICK TREE FARM
Coram, Montana

ACREAGE
170 acres in northwestern Montana, Sec. 20 T31N, R19W;
purchased in 1972 by Vicki and Bill Byrd-Rinick

MANAGEMENT OBJECTIVE
"We're logging stands of dense, decadent lodgepole pine to make a more aesthetically pleasing and merchantable stand for the future."

GEOGRAPHY
3,100 feet elevation, southwest-facing aspect, sandy-gravel soils

CLIMATE
90-day frost-free growing season
24 inches annual precipitation

STANDING TIMBER VOLUME
Standing volume in 1998: 20 to 30 tons per acre (approximately 850 mbf total; 5 mbf per acre)

STANDING TIMBER SPECIES
Lodgepole pine: 85 percent of total stems
Douglas-fir: 10 percent
Others: 5 percent
Average tree age: 70 years
Average tree size: 8 to 10 inches dbh
Average growth rate: 1/10th to 1/15th inch diameter per year Number of canopy stories: 2

UNDERSTORY VEGETATION SPECIES
Primarily timothy and beargrass with duff, interspersed with kinnikinnick, huckleberry, glacier lily, violet, ladyslipper, trillium, Indian paintbrush, oxeye daisy

WILDLIFE DIVERSITY
Mammal sightings: Columbia ground squirrel, white-tailed deer, black bear, elk, moose, grizzly bear, mule deer

Bird sightings: robin, black-capped chickadee, ruffed grouse, common nighthawk, tree swallow, dipper, varied thrush, Townsend's solitaire, cedar waxwing, junco, sparrow, pine siskin, osprey

HARVESTING HISTORY
1984: selective clearcut on 55 acres (conventional logger)

1985-87: thinning on 20 acres (horse logger)

1995-96: single tree and group selection on 40 acres (Bob Love); removed 10 trees per acre (2,500 bf per acre, 40 percent of standing volume)

ECONOMIC RETURNS
Average gross income from harvesting: $9,600 total in 1995-96 harvest

$240 per acre for 40 acres

$220/mbf for merchantable sawlogs (50 percent of total harvest volume; other 50 percent unmerchantable)

Costs are estimated as 100 percent of gross income; a goods-for-services contract allowed the logger to keep all proceeds to fund precommercial thinning in the same stand.

LOVE'S FORESTRY
Bob Love works as an independent consultant forester and logger near Kalispell, Montana. He uses a 26-inch chainsaw, a 60-hp tractor-skidder, and sometimes a portable chipper and sawmill. What makes Love different from other "gyppo" or independent loggers? Love:

- works as a both a forester and a logger, planning and carrying out timber sales

- works for an hourly wage or under a "goods for services" contract instead of being paid per volume logged, so he has no incentive to overlog

- maintains the integrity of the original forest by using as much of the tree as possible to reduce the number of trees he needs to harvest; by adjusting his harvests on a micro-scale, from acre to acre, to maintain historical tree species, native plants and wildlife habitat; and by considering all the reasons not to harvest a tree before he cuts one down

- specializes in logging forests that have been poorly logged before, or dense forests that don't have much merchantable timber, to improve the quality of the forest for aesthetics, recreation and future timber harvesting, and to reduce wildfire and disease risk (for example, on the Byrd-Rinicks' property, lodgepole has grown into thickets that slow the growth of individual trees and increase fire hazard, so forester Bob Love has removed tons of unmerchantable timber to increase growth opportunities for more merchantable timber; Love can only give the Byrd-Rinicks a gross return one and one-half times that of Goebel's (on a per-acre basis) after removing more than five times the volume because most of the lodgepole on the Byrd-Rinicks' property are unmerchantable)

"The good [logger] loves the board before it becomes a table, loves the tree before it yields the board, loves the forest before it gives up the tree."

—Farmer and writer Wendell Berry, *Preserving Wilderness*

Logging started in the northern Rockies in the late 1800s to supply mining timbers and railroad ties to the booming West. Early photographs show logging as crude work. The loggers, facing large trees with small machinery, needed to pry and shove wood out of the forest.

"Back then we really worked the woods," said 87-year-old Donald McKenzie in an oral history transcription housed at the University of Montana. "We were dealing with stubborn horses or mules to skid the logs and we'd have to pay a lot of attention to our work." If streams were near, loggers dragged their logs by horses or mules to a water flume to shoot the logs down into the river. Other times they dragged the logs straight along roads or slid them along snow trails, using horses, oxen, or mules. Later, the loggers reeled the logs in with clunky "steam donkeys." Sawyers clipped cut logs to cables strung through the forest. The cables ran from the forest to a steam-powered, rotating drum. Once the logs were attached, the steam donkey fired up the rotating drum and reeled in the logs.

Decades ago, when there were still giant trees to cut, loggers would commonly cut a notch in a tree and put a wooden plank or "springboard" in the notch, then climb on it and saw above the tree's thick, flaring butt. Working as partners, the loggers would pull a big crosscut saw back and forth and gun the tree down. Then they tied the butt end of the log to a yoke and guided a team, typically of three horses, to a water flume to unload the log. In the river, loggers with pike poles and spiked boots jumped from log to log to keep them flowing downstream to the mill. "Those river men were proud," McKenzie said, "but they had cruel work. Their wool trousers would get waterlogged and their sharp boot studs would dull. Sometimes, despite the loggers' best efforts, the logs would seize into mounds so tangled it would take weeks to work them apart. Everything shut down when the logs jammed, and those times I wished the men could just find a string and yank the piles apart."

The whole logging crew lived together out in the woods, and their days were a constant rhythm of sawing and moving trees. Nights brought a respite of stories, card games, and fiddle music. Six days a week, loggers worked from five in the morning to three or four in the afternoon. Then at spring breakup, when mud suspended logging for a month or more, the workers got a year's wages in gold coin and celebrated with a big drunk in town.

"Logging was not just a career but a whole life decision," said McKenzie. "Once you chose to be a logger, you would spend your entire life out in the woods." McKenzie still remembers a few loggers he met in the company camps.

They included Rose, an effeminate French-Canadian who only knew swear words in English, and Buster, a blocky (and lucky) boy who was missed by a falling tree one month, then survived a truck crash off a bridge the next. He also remembers how, after hours of muscle work, he looked forward to warm bread and beans in camp. "Those camps were somethin' else," McKenzie said. "My friends were there, and my bed and food. After boys started logging, those camps were the only homes they knew." Old loggers would often choose pension work in the camps, such as sharpening saws or feeding horses, over retirement in town.

In the early 1900s, a healthy team of two loggers could cut 2,000 board feet of timber—three or four big trees—in a day. Then it would take anywhere from two days to four weeks to move the wood by animal and water to the mill. Now, with mechanization, one logger can cut more than 10,000 board feet, delimb it, skid it out to the landing, and haul it to the mill the next day.

"Back then," McKenzie said, "we could only afford to take the big trees, the stuff you'd have to stretch your arms to get around. Now we can take almost all the trees."

Zen and the Art of Logging

"Logging today is like lighting a match," said Art Partridge, a retired forestry professor from the University of Idaho. "You can either start a warm fire or burn something down." Careless logging marks land for decades—a lush forest withering dry, a road rutting up a hillside, skinny trees left bending down in snow or snapping off with wind. But careful logging leaves stands so pure, you have to hunt for stumps.

A half-hour drive west of Columbia Falls, Montana, lies a road that bear cubs cross and bull moose swagger along. A few houses stand along the road, spaced far apart, and most fence posts carry timber-industry stickers, such as "We support the timber industry" or "Hug a logger and you'll never go back to trees again."

Between mailboxes numbered 4995 and 4885, one box reads simply "Love." Bob Love worked almost 20 years felling trees throughout the West for Plum Creek Timber Company. He used to hook up cut trees to jammers—old, home-made machines with a set of cables running from an engine on the back of a truck out into the forest. One of the best jammer operators he ever saw, he said, was a 70-year-old he met at a bar. The man had set choke on jammers his whole life and worked "like a magician." Love now works as an independent logger and forestry consultant for landowners.

I first met Love at a forestry meeting in Kalispell, Montana. He wore a "No Fear" T-shirt and Levis, and he had patient eyes and workingman's hands—hands like tools, more steel than skin. We talked during the morning break.

METHODS FOR TREATING SLASH

Conventional method of disposing of lodgepole slash in giant piles

Lodgepole slash on the logging deck on David and Kathryn Owen's property. The slash is divided into neat piles of merchantable timber, firewood, and slash to be burned.

Bob Love chips onto slash trails debris that might otherwise have been burned.

Logging, Love told me, is not about boots and machinery, or about board feet and dollars. "My logging is Zen work," he said, close to me. "It's measured not by what I've done to the forest, but by what I haven't done; it's not what I've taken that matters, but what I've left."

Weeks later, Love and I went out to one of his logging jobs and chipped slash. Love uses a 26-inch Stihl chain saw to cut trees and a John Deere 450 tractor—a medium-sized machine with a 50- to 100-horsepower engine—to skid the trees to a deck. There he delimbs them and bucks them into 16-foot lengths. Some other "gyppo," or independent, loggers use this same equipment, but Love works his tools gently, leaving the land unmarred.

Love assigned me to the drum chipper to stuff tree limbs into a spinning, bladed wheel. The chipper pulled the wood in quickly and threw out chips with a sound almost as loud as an airplane engine. For hours that day, I smelled only cut wood and heard only dull machinery noise. By chipping limbs after harvesting, Love explained, he can cover machinery trails and keep nutrients on-site while also meeting state standards for residue disposal to control fire hazard.

Landowners use slash—the remains or "guts" left after delimbing and bucking—in various ways. To reduce cost, most loggers machine-pile the slash into truck-sized mounds and burn it during the spring or fall after harvesting, sometimes creating conflagrations that sterilize the soil for years. Others scatter small limbs to recharge the soil with nutrients and burn the tops in small piles, use them for firewood, or sell them to pulp or post-and-pole operations.

Sometimes Love will scatter small slash and burn the large debris. But he increasingly chips the unusable debris. Chipping isn't perfect, particularly when it's spread thick because it can retard decomposition and cover growing plants. It's also much more expensive than burning slash piles, but about the same price as broadcast burning.

But when chipping is done right, just a thin layer covering skid trails, it's window dressing for the logged forest. When we finished, woolly trails ran through the forest, instead of dusty roads and landings. "It's as though you healed the logging wounds," said the landowner.

While I finished chipping, Love cleaned up an area near the log house. He drop-started his saw to cut large pieces of slash into manageable sections and moved the blade casually, as if it were part of his hand. After a few minutes, he had split Douglas-fir limbs into fireplace logs and had transformed a lodgepole top into a long fence pole. Most loggers would have burned these small pieces of wood, but Love uses the whole tree as good hunters use the whole animal.

We stopped twice on the way back to Love's that evening. First, we looked at a landowner's property along the highway, just 20 minutes from Love's house. The place looked like a junkyard, full of dusty roads and scraps of trees. We walked up into it and stood next to a 20-foot-circumference burn pile of tree debris.

"I talked to these loggers," Love told me, "and they said they did a professional job. They told me they had only removed the dead and old trees and made a little money for the landowner. But it doesn't take an expert to tell a good logging job. You can tell by whether the land still looks and feels like a forest." My nose filled with heat from the bare soil as he talked. I remember talking to a landowner in Georgia who told me, "People say looks aren't everything, but that's because they don't look closely at the forest."

I said to Love that I hated the looks of this place, but I sympathized with the loggers. I told Love about a time I cut lodgepole pine and Douglas-fir in a mixed forest. I started the saw and made a face cut in the direction I wanted to fell the tree and a back cut behind it. The tree creaked forward, then slammed down like a thunderclap. I paused for a second, then cranked up the saw and cut another tree. Sometimes the saw stuck, bit down on itself, but I just drew it out and forced it in again with my hips and shoulders. Nothing stopped me — I had cut a large tree, and I cut a larger one after that, getting all the best trees. My arms shook with adrenaline. Finally, I put down my saw, took off my gloves, and looked around at the strong trees, now limp on the ground. These were trophy kills.

I cut few trees that morning, but I had cut all the wrong ones: the big rather than the small. It's not necessarily wrong to cut the big trees; after all, that is where the landowners eventually make their money. But I was impatient and cut the trees before they were ripe — before their growth had appreciably slowed and neighboring trees had grown up to replace them.

Love and I stood there silently after I told him the story. After a few minutes, we turned around — not in a planned way but with one starting to turn and the other following the momentum — and we walked back to the truck. "I'm going to show you what I do," he said.

We drove down the road, then stopped at a different property harvested two years ago, about the same time as the scraped-up land where we had just been.

"Let's go," he said. Coming into this forest from the highway was like coming into a room; I moved from asphalt into trees as abruptly as if I had walked through a door. A matrix of light and air diffused beneath the canopy.

Love's dog, Murphy, and I followed him and stopped on a forested hill. "That's one of my skid trails," Love said, pointing to a barely noticeable indentation below us.

When Love first came to these forty acres, he spent a full day just walking it. He listened for birds and noted the sizes of trees that different bird species were using for nests. He looked for animal tracks, which told him where the animals fed and traveled. Ospreys, herons, and kingfishers hunted in the pond, red squirrels nested in limbs, and snowshoe hare hid in dog-hair thickets to elude predatory goshawks and owls. Love looked for old stumps that told him when the area was last harvested and which tree species had grown there.

"Now, look over here," he said, crouching down and pointing at the south slope across from us. The previously thinned stand of Douglas-fir and western larch showed the dry, brown vegetation beneath them. "And look up here." He twisted around to a patch of lodgepole we had walked through. I remember it was thinned, but still so dense in places that I had to scrunch my shoulders together to get through.

"Now most loggers would have done this the other way around," he told me. "They would have gotten rid of the lodgepole pine and only harvested lightly on the drier, southwestern slope. But I saw that shade-intolerant larch grows on the drier slope and needs more space and that deer and elk bed in the lodgepole and need cover. Also, lodgepole grows up from one disturbance, so it's all about the same age. This stand is still young, about fifty years old, and has at least another twenty-five years before it becomes susceptible to wind, bark beetles, or fire. I'm not following forestry textbooks here, but I'm paying attention to what the land says to me. I'm paying attention to the microvariations that make one piece of land different from another."

Love and I moved down into the forest. We walked as friends do, easy with each other. We both slowed and stopped at a place darkened by plants and trees.

"A cut is a serious move, with all that life in your hands," he stressed to me. "But I really can't give any guidelines for selecting trees to harvest any more than I can give rules for treating people right. It's all based on conscience, and having to choose trees is as hard as choosing between friends." Love removes mainly scrawny, crippled trees. But before he removes any, he makes a quick mental list of reasons not to cut the tree down. "You need to think of all the reasons why you shouldn't cut down a tree," he told me. "You need to think about what that particular tree in that place does for the soil, the vegetation, and the wildlife, compared to other trees around it."

Much of the land Love works in the Flathead River Valley was burned severely in the Halfmoon fire of 1929 and again one hundred years before that. He logs in a pattern that mimics the burns. He saves dominant trees that have bark thick enough to resist fire and thins out small competing trees, while

TECHNICAL NOTE

Which Trees Should I Cut?

▶ Why Should I Harvest Trees in the First Place?

More than half of the Nonindustrial Private Forest (NIPF) owners in the country will harvest timber sometime during their land tenure. Landowners who selectively harvest timber do so to bring in supplementary income or to improve timber quality for the future by reducing competition between trees and reducing the susceptibility of the trees to insects, diseases, fire, and wind.

Forests naturally go through stages of crowding and thinning, as individual trees constantly die and others grow in their place. Selection harvesting can emulate this natural process.

▶ Do I Need a Timber-Harvesting Plan?

Landowners should not just harvest helter-skelter. Each landowner needs to develop a management plan to identify what he or she really wants and how to get there without damaging the forest. Your plan will help you avoid expensive mistakes, prioritize your expenditures, and qualify you for government cost-share programs. In some states, a timber-harvesting plan is required by law before the first tree can be cut.

Courses and books offered by state forestry and extension services can help landowners write a forest management plan. You should attend as many courses and read as much as you can to understand your forest. However, you should go to a forestry consultant to have a plan professionally written. The consultant can also work with you to mark timber and choose a logger.

▶ When Should I Harvest?

Harvest when the stand is too dense. Harvesting can keep the forest from stunting and becoming infested by disease and fire (the natural thinning mechanisms). You can use tree spacing to roughly judge, or annual growth rates for a more accurate assessment. If you use spacing to gauge density, a forest may be too dense when trees are less than 10 feet apart on average. Allow tighter spacing for shade-tolerant species such as lodgepole pine and wider spacing for shade-intolerant species such as ponderosa pine. You can also assess density by measuring diameter growth from year to year on specific crop trees and seeing whether the rate of growth has continued or has stagnated, or by seeing if the crowns are touching.

▶ How Much Should I Harvest?

The growth rate should generally be the ceiling for your harvest volume; this way you are harvesting the interest from the forest, not the principal. You can determine this figure by measuring growth rates for individual trees (typically 0.1 to 0.5 inch per year), then going to charts available from state forestry offices that translate inches of growth to board feet.

You may need to harvest above the growth rate, perhaps by harvesting trees in small groups, if you have a logged-over or stagnated forest; if you have a stand where more than 10 percent of the tree volume has been killed by insects, disease, wind, or fire; or if you want to take advantage of a lucrative timber market.

▶ Which Trees Should I Harvest?

- Try to remove dead, diseased, damaged, deformed, and suppressed lower-canopy trees over healthy, upper-canopy trees, although you must leave some snags for wildlife and down woody debris for soil nutrients.

- Try to mix in some "sweets" with the "sours" by removing some higher-quality trees to make the harvest profitable. But do not harvest so many of these dominants as to restrict opportunities for future harvesting. Harvest mature trees that have slowing growth and have neighboring trees that can replace them.

- Mark with survey tape the trees you wish to take in a portion of your property and then pretend those are gone. How does your forest look? Does it still have a variety of tree species and sizes?

leaving some of the wood on the ground for nutrient recharge. He cuts down patches of small trees that act as fire tinder, although he retains some of the dense growth for wildlife cover. Love prunes maple and serviceberry in the understory to encourage green browse, for example, but leaves some thickets for bird nests. This is Love's Zen logging—it concentrates on what's not done. In a pamphlet, Love writes:

> Like a wolf who spots the cripples in a caribou herd, you must identify a surplus that can be removed without harming the integrity and nature of the pre-settlement forest. [You] must respect the dominant trees in the forest as mature and battle-tested bucks to be saved.

On most properties, Love removes about one load of timber every 10 acres, about 300 to 400 bf per acre with a gross gain of $120 per acre on average, because of low profits from skinny lodgepole. Love charges $35 to $70 per acre for his services (2000 prices).

Admittedly, Love is lucky. Most of the forest owners in his area do not depend on the land for their incomes. "A lot of people here can choose profitable harvests over profit-maximizing harvests," Love said. "And a lot of

A stand of lodgepole pine logged by Bob Love

people here are willing to forego a profitable harvest in the beginning so I can fix up their land and create some places for merchantable timber to grow for the future."

Landowners who bring income to the land are increasingly common in the West, improving the opportunities for Love's maverick logging. Along his road and up on Big Mountain where Love has worked, neighbors have come to him, not even inquiring about cost, but just wondering if he can do to their forest what he has done to other forests in the valley.

TECHNICAL NOTE

What Is Sustainable Forestry?

"The major principle of sustainable forestry is to remove trees which will not survive, taking those nature has already selected against," said Orville Camp, a western Oregon landowner who has written a book on sustainable forestry. "You sustain the forest ecosystem by retaining the larger trees and cutting away the smaller, weaker trees."

The larger trees are harvested when their growth slows and other trees are ready to replace them in the canopy. "Be careful which big trees you remove," Camp cautions. "Say you have two trees growing right next to one another. A conventional forester would say automatically that you need to cut one of them because it'll interfere with the other one's growth. But the trees could be growing just fine, and sometimes the best trees grow in clumps over rich soil. You need to compare the trees to one another in the forest so you're maximizing growth on each tree."

The landowners in this book follow many practices of sustainable forestry in addition to Camp's fundamentals. These landowners:

- Typically harvest within the annual growth rate
- Use treetops for posts and poles or firewood, or burn them in small piles
- Retain fallen deadwood to reduce water runoff, recharge the soil with nutrients, provide habitat for small animals, and nurse seedlings
- Retain standing dead trees, called snags, for owls, woodpeckers, and secondary nesters
- Harvest diseased trees to prevent future infestations
- Use prescribed fire in appropriate forest types

- Protect wetlands, ponds, streams, and rivers from harvesting (preserving about 100 feet on each side of primary streams)
- Shrink the size and reduce the number of roads and skid trails
- Fortify roads against erosion with dips, water bars, and culverts
- Minimize soil disturbance by using small skidders, winching trees to skidders, and concentrating harvesting in the winter, when the ground is frozen
- Maintain or restore tree and plant species diversity and protect rare plants and animals now on the property
- Preserve old-growth areas in the forest—roughly those with trees more than 100 years old

How do you know whether you are practicing sustainable forestry? Ecological monitoring is one important gauge, says Richard Hart, an Oregon ecological consultant who spent his youth as a monk and his early adult years as a Porsche-driving businessman.

Hart became a monk at 14 and was soon put in charge of monitoring a watershed for an abbey. At first, he thought it a simple job of monitoring flow rates. Then his mentor, an 80-year-old monk, showed him how to look at the soil and meander through a forest without creating a trail. "Soil glues the forest together," said Hart, peering up from behind his glasses. "An undisturbed forest soil has a web of fungi and an airy texture like Wonder Bread. It powers the whole forest." Hart later traveled through Europe on a bike and looked at how ancient abbeys planned ahead so they could forever use their forests, employing such methods as marking some trees to be harvested a century later to replace roof timbers.

Monitoring the forest is crucial to test whether the assumptions of sustainable forestry are valid, says Hart. The first step is to look for good spots for monitoring, such as watersheds, heavily and lightly harvested areas, and unharvested areas. Permanent steel stakes should be driven into the monitoring spots, and a variety of parameters should be measured: soil moisture, soil type, soil compaction, ground light, plant populations, wildlife signs, and fallen and standing dead trees, as well as tree age classes, species, heights, and diameters. Photos in multiple directions should also be taken at each plot. All of this can be done without expensive equipment, Hart points out. Professionals, however, may be needed to help determine whether the changes in these factors actually diminish the forest's integrity.

Love and I drove to his house. He lives near the North Fork of the Flathead River just upstream from where the riverbed widens, where the mountains of friable rock partly give in and melt down in alluvial fans.

Love's house stretches out in one long level. A garden lies in front, filled with asparagus, onions, beans, and lettuce. I felt like letting my mouth loll open and tipping in all the fresh vegetables with their delicious and watery flavors.

After I showered and changed out of my field clothes, Love took me to a tepee outside where he and his wife Inez lived for five years before building this house. The canvas skin of the teepee stretched across lodgepole pine poles. "This is a Sioux-Cheyenne design," he said. "It's roomier than those of the Blackfeet." Love explained that he erected the tepee by setting up three thick poles and smaller poles between them, then wrapping the skin around the main tripod.

The winters were the hardest part of living in the tepee, Inez said. She hated to get up and change clothes in the icy air, then crunch through a mile of snow to get the car for mail or groceries. But she would never give up the quiet nights at the tepee for comfort while she was still young. "The world was so dark and silent those nights," she said. "It seemed as though my voice would crack the sky."

After moving out of the tepee fifteen years ago to start a family, Bob slowly expanded on a small cabin he had built near Glacier National Park, making it into the house where they now live. When I looked at the house carefully, I saw the rooms as jigsaw puzzle pieces, each representing a stage of construction. I slept in the first room Bob built, just a simple bedroom with a heavy plank door, chinked log walls, and a makeshift kitchen with a hand-pump in a corner. After this room, he built the master bedroom—a sanctuary with a wide ponderosa pine-hewn bed, floor spreads made from deer hides, and a series of long windows, each a sequence of the stream behind the house. Then he added a couple of bedrooms for his children and a modern kitchen. Love still cooks pancakes on a wood stove in the house every Sunday, using 130-year-old sourdough starter, just as he did when he had the little cabin.

For dinner, Inez made a spicy bean mixture and homemade tortillas, and I filled myself with them as we talked. Inez told me she teaches at a local elementary school. She cares personally about the students. One weekend, for example, she and Bob took a couple of sixth graders from town on a trip to the Museum of the Rockies in Bozeman to help them research a report they were writing for school. The children were comforted by Inez's soft demeanor and entranced by Bob's clear explanations at the museum. "They were nice sixth graders," said Inez, who raised two children with Bob. "They said thank-you almost as many times as they told a dirty joke."

Conventional uneven-aged logging in lodgepole pine in Montana

Later, Inez served warm blueberry crisp. Bob and I forked through the dessert and talked about forestry. "When I was young and went into the forest, I would look only at the biggest trees; they catch everyone's attention," Bob said. "Now, I look at the forest as an 'interbeing' where soils, plants, and animals all interact.

"I'll give you an example. If you go out to see other loggers' work, you'll notice that a lot of them don't care what their equipment does, as long as it gets the trees out. Sometimes their tires will make furrows 3 feet deep like they're planting potatoes. But I lay out piles of limbs where my skidder turns, because I realize that the soil affects water and plants and wildlife and trees." Bob leaned into the table, closer to me. "I dream of other loggers doing little things like that to fix up the land, then having bull trout swim back into their streams, mule deer jump back into their meadows, and black bear crawl back into their hills."

Love's logging in lodgepole pine in Montana

The next day, I looked out back and saw logs that could go into all sorts of products—house frames, boat hulls, tables and chairs, posts and poles, cardboard boxes and notebook paper. I asked Love where his wood went.

He brought me behind the house and started a portable sawmill in the cold morning, yanking the string on an engine again and again. It finally caught, puttering and shaking, gooping oil and snorting smoke. He pulled a lever, and a log he had lifted with a tractor came down into a long steel carriage. Love locked the log into place on steel arms, then he went to the saw, methodically fixing the alignment with knobs and cranks. I enjoyed watching him working quietly behind the saw, his hands acting from memory. Love made the last adjustments, then he pointed to the sawdust below the portable Wood-Mizer Mill. He told me that he leaves most of the sawdust on the ground now so the soil doesn't freeze and buck the saw.

He finally shifted the drive bar into place, and the angry saw — one 2-foot blade spinning on the side, two 6-inch edger blades whirring on the top and bottom — moved through the lodgepole. Sawdust spit out from a hood on top, and three bark sides fell off, which would be sold for siding or firewood. When the saw returned, Love adjusted it to cut a 2-by-4. In a commercial mill, a saw first reduces the log to a square cant by cutting all four sides of bark, then cuts lumber from the cant with sharp, laser-guided blades. Here the saw just moves from one end to another, back and forth. Love alternated levers for the vertical and horizontal blades and worked the average-sized tree, probably 400 board feet, into three 2-by-4s and some scrap bark slabs. I helped him stack the boards into a neat pile near the mill, and together we covered the pile with a tarp to keep it from warping. The white-and-pink wood smelled so strong I could almost taste it.

Love mills a quarter of the wood he harvests and sells the rest to American Timber, a small mill in Olney, Montana, near the Canadian border. The mill looks like it has for decades — a metal shed full of machines and conveyor belts. But as the average size of timber has decreased over the past decade due to overlogging of big trees and shortening of time before trees are taken, the mill's efficiency has increased. The mill now uses lodgepole pine with diameters at their tops as small as 4 inches rather than 6 inches, the standard diameter during the early 1990s. The mill also gets about 25 percent of its wood from NIPF owners, compared to less than 10 percent a decade ago when it still used primarily timber from national forests.

A few weeks after visiting Bob Love, I visited a couple he had recently worked for, Vicki and Bill Byrd-Rinick, who live just a half-hour's drive west of Love's house. They invited me in as soon as I arrived — Vicki wearing sandals and sweats and Bill in a button-down shirt and jeans. I sat down to carrot cake and coffee with them at the kitchen table. Snow blew and wrapped around the banks of the Flathead River below like a scarf.

"We have some stories to tell," said Vicki. Three different loggers — a conventional logger, a horse logger, and Bob Love — have worked their land over the past fifteen years.

The Byrd-Rinicks first started logging in the mid-1980s, about a decade after they purchased their land, because of an insect epidemic. Mountain pine bark beetles had turned their green lodgepole forests psychedelic red. Lodgepole pine commonly goes through a 70- to 100-year cycle, during which it becomes susceptible to mountain pine beetle, then becomes vulnerable to wind blow-down, dwarf mistletoe infestation, and crown fires.

The first logger the Byrd-Rinicks hired came at the recommendation of the state forester and was profiled in the state forestry magazine. "This guy was supposed to be an expert," Bill recalled. He came with shiny equipment — a feller-buncher with two arms that vise around the tree while a saw swings out from the bottom; a grapple-skidder that clamps over as many as eight full-sized trees to drag them away; and a Swedish processor that reduces trees to logs in minutes. The logger assured Bill that his system was very advanced, with no wood waste or soil damage. But the logger worked the forest while the soil was still wet, tattooing the ground with tire marks, and he left giant, tangled piles of slash.

■ THE PROCESS OF PLYWOOD

I traveled to a plywood plant to see how plywood is manufactured, to see what happens to some of the big trees—trees with top diameters greater than 7 inches and breast height diameters greater than 16 inches.

Next to furniture veneer, plywood is the most valuable coniferous wood product. Mill prices in 1999 for 16-foot-long, plywood-diameter Douglas-fir and western larch logs surpassed $550 per mbf, about $150 more per volume than smaller trees of the same species. "Plywood and stud lumber are scaled differently, so it's hard to compare, but plywood is substantially more valuable per unit volume," said Jeff Webber, Stimson Lumber Company's plywood plant manager in Bonner, near Missoula, Montana. Plywood is as much a mechanical process as a natural product. "Plywood's not really made from wood," one old forester told me, "It's made from machines and glue and such."

Imagine the wood that goes through the plywood plant as an apple. It's heated, then peeled, then sliced in thin layers down to the core. The slices are dried, spread with glue, and placed crosswise one on top of the other in layers, like a sandwich, and then the layers are pressed together in a compressor.

First, a toothed ring debarker gnaws the bark off the tree and saws cut the logs into 8-foot units. Conveyor belts carry the naked trunks into vats, where they soak in hot water for about ten hours. Soda added to the water keeps the pH neutralized; without the soda the natural acidity of the conifers would turn the water into a caustic soak that would eat apart the wood. The tree cores heat up to around 100 degrees Fahrenheit, so they neither mush nor fracture on the lathe. Steel arms center the log on the lathe and lift it, then a knifed carriage settles down, cutting off 0.1- and 0.125-inch-thick sheets from Douglas-fir and western larch, and thinner 0.17- and 0.19-inch-thick sheets from grand fir and western hemlock. The veneer comes off steaming and gorgeous. It's diaphanous and looks like something that would dissolve in pieces in your mouth like a church wafer.

The translucent wood moves along conveyor belts and rollers at 1,100 feet per minute. It is cut into various widths, depending on the size of the plywood being manufactured—an average tree can yield almost twenty 8- by 4-foot sheets. Scanners clip out defects and sort the sheets into groups by moisture content—super sap, sap, or heart. Wood from a Douglas-fir that grew by a stream, for example, would have a higher moisture content than an old western larch. Then the sheets move into ovens set at 400 degrees Fahrenheit, with the oven time adjusted according to the wood's moisture content. The sheets usually dry after just a few minutes, but each piece is monitored so water does not build up and blow into steam in the hot press, which would pop the other plywood layers apart. In the layup line, mechanical arms extrude glue on each wood sheet—a secret mixture of resin, soda, and flour, with animal blood as a foaming agent—then lay another sheet cross-grain over the first. With a hydraulic expiration and groan, the 4-, 6-, or 8-ply sheets hot-press together.

Down below, in the basement, a recycling process runs alongside the main processing up above. Vibrating screens sort out shavings by size—some thick pieces called overs, some regulars, others fines—and belts send them off to be used as boiler fuel or paper pulp. Bark and cores from the lathe are reworked into beauty bark and landscape timbers.

Up above, the raw plywood sheets still need to be finished. Pluggers fill knots or holes with solid wood. Ethylene nozzles lacquer down rough faces. A saw squares the sides. This, finally, is plywood: the strongest wood, one that does not warp or split or crack.

"I wrote to the logger time and again to ask him to at least dispose of the slash," Bill said. "Finally, after three years, he burned it. He called later and told me that a slash fire had gotten a little out of hand." The fire had burned madly, a mix of flames and gray-white smoke billowing into the adjacent national forest. A month later, Bill received a bill for $2,200 from the Forest Service for controlling the blaze. "The contract saved me," Bill said. "I highlighted the

EQUIPMENT SMALL LANDOWNERS
MIGHT USE IN LOGGING

A 21-inch chain saw and wedges to help in directional felling

A cable skidder on a small track-type tractor. The cables come off the back of the tractor. A plug on the end of each cable fits into a sliding housing, and the cable cinches around a log. The tractor then pulls in the cable and drags the log to a landing or road.

TECHNICAL NOTE

Foresters, Loggers, and Logging Contracts

▶ How Do I Find a Good Forester?

Landowners often first choose a good forestry consultant. Bob Love is an unusual case, being trained both as a forester and a logger. Some sly loggers tell landowners they don't need a "middleman" forester. But doing a forest harvest without a forester is like building a skyscraper without an architect, and letting the contractor do all of the planning and design on the cheap. Everything may look all right, but the structure of the forest is compromised. A forestry consultant will help you write a management plan, set up a timber harvest, develop a contract, and choose a logger.

Landowners can find good foresters by getting recommendations from neighboring landowners, state foresters, extension services, and private organizations, such as the Forest Stewards Guild, which has a database of almost four hundred foresters, natural resource consultants, and loggers like Love committed to sustainable forestry. Landowners need to find a number of possibilities, do careful interviews, and drive out to forests to check references. Most landowners say that choosing between foresters is a decision they ultimately feel in their bones.

▶ How Do I Choose a Good Logger?

"There are rational steps landowners can take to make sure they have chosen a good logger," says Patrick Heffernan of the Montana Logging Association. Once you decide on a forester, and decide to harvest timber, you and the forester should go through these steps to get a good logger:

- Gather multiple references from the bidding loggers. Visit their work sites and talk to the landowners who hired them. Remember that you want your logged forest to still look beautiful and feel natural, so find a logger who has left the forest this way. Try to find similar properties in terms of tree species and annual growth so you're not comparing a wetter, more productive forest to a drier one.

- Choose a logger on the basis of bid price and past performance, not necessarily the equipment used. A good logger can work well with any equipment, though there are trade-offs with different types. Larger equipment moves wood at lower costs because it can cut and move fast, but it demands big roads, and its owners must harvest high volumes to pay the equipment costs. Horses are slower and, thus, more expensive, but they demand less space and leave smaller skid trails. Small mechanical equipment, such as Love uses, has some of the advantages of each—smaller trails like horses and lower logging costs like large equipment—but it's not particularly specialized for small or large volume harvests.

- Look for loggers who are experienced but still innovative, and who have gone through certification or stewardship programs.

- Ask the loggers reflective questions, such as why they got into logging and what their biggest successes and failures in logging have been.

- Carefully examine the contract from the chosen logger to make sure it includes all the necessary provisions.

- Secure a performance bond.

- Meet on-site with the logger to discuss harvesting restrictions and answer any remaining questions, then monitor logging operations at least once a week.

▶ How Do I Prepare a Timber Sale?

- Do a thorough inventory to determine the real value of timber on your property.

- Talk with adjacent landowners and notify them about your harvesting plans. Ask if they are willing to share road-building costs or are interested in a combined timber sale.

- Mark the timber to be harvested and mark property boundaries.

- Write a description of the timber sale. Include the acreage, timber species, and average size; estimated total volume; special restrictions such as damages to residual trees, snag retention, and streamside setbacks; management objectives; vision for the condition of the forest after logging; dates for the harvesting; dates and times the sale will be shown; directions to the sale property; your name, address, and phone; and deadline for bids.

- Send the sale description and bid request out to appropriate newspapers and magazines, as well as to specific loggers you have heard about. Compare the bids to your own price estimates from the inventory.

If you have a commercial sale, the logger can pay you in one of two ways:

- a lump-sum sale, or stumpage, where the logger pays you a single fee after the contract is signed, but before logging begins. This method is the most common.

- a mill-tally sale, where the logger pays you per portion of wood sold, minus prearranged logging costs. This method is based on load receipts from the mill.

Both of these methods are reliable, as long as you have done a thorough inventory beforehand so you know the value of your timber. With either payment plan, the logger should pay you at least one-third to one-half of the gross income.

If you do precommercial thinning, you can pay the logger in one of two ways:

- per acre or hour of labor, after talking with a consultant or state forester to determine a reasonable rate.

- with a goods-for-services exchange (used with Love here and with Nathan Arno in Chapter 1), where you allow the logger to harvest a volume of merchantable timber goods in return for precommercial thinning or prescribed burning services.

▶ What's in a Timber Contract?

Contracts between landowners and loggers are required by law if the timber sale involves more than $500 in goods or services. (See Appendix II for a sample timber harvest contract.) "Good contracts are clear, without much legal jargon," said Scott Burnham, assistant professor at the University of Montana law school, "and are still foresighted enough to cover potential problems." At a minimum, timber contracts should:

- Clarify the logger's role as independent contractor

- Guarantee the logger's coverage under comprehensive general liability and workers' compensation insurance
- Define logging-area boundaries
- Define the time of year and length of work period for logging
- Define tree-marking guidelines, such as size, species, and quality (e.g. deformed) of trees to be harvested, along with total and per-acre harvest volumes, though you might be better off marking the trees yourself or hiring a consultant forester to do so
- Explain road-building standards, slash disposal requirements, retention of snags and down wood, riparian-area restrictions, and soil-moisture restrictions
- Establish a method and schedule for payments from the logger, or from the landowner in a precommercial sale
- Set a performance bond from the logger of up to one-third of the total value of the sale, to be held by the landowner in escrow until all the contract requirements, including slash clean-up, are fulfilled
- Release the landowner from liability for all injury and damage from forest hazards, and from violation of laws and regulations pertinent to the logging operation (indemnification)
- Establish penalties for violating the contract and means for resolving disputes, such as attempting mediation before litigation
- Gain the witness and seal of a notary public

part saying the logger was responsible for fires on the logging site, and I sent it to the Forest Service. I haven't heard from them, or the logger, since."

"When I first hired the logger in the 1980s, I took him out to the property, and we walked through the sale—so he knew what I wanted," Bill explained. "I gave him a relatively simple contract that spelled out the area he had to work with and confirmed that I would receive scaling tickets from the mill. I also put all the liability for timber treatments on the logger."

Bill took in a breath. "I'm glad I put what I did in the contract, but I should've added some things for sure—even just a performance bond, so I had some money to pay for things he screwed up."

Five years after the conventional logger had worked on their eighty acres up the road, the Byrd-Rinicks decided to harvest the ninety acres they lived on. "We wanted to be more gentle on the land this time," Vicki said, "so we hired a horse logger." The logger was a patient man who worked Belgians, a team that could pull a 30-inch-diameter tree.

Still he ended up leaving mounds of slash, holes in the forest like gaps in teeth, and landings infested with weeds from the horses' hay. "We trusted the logger more than we should have just because he worked with horses," Vicki confessed. "It was really the same mistake we made before in not checking up on references and not monitoring the logging operations."

Last spring, I visited Jim Lotan, director of the Northern Rocky Mountain division of the North American Horse and Mule Loggers Association. I met Lotan in front of his house in the Bitterroot Valley in Montana, and he looked like a gnome, with wizened eyes and an arrow-point goatee. We went into the house where he showed me slides of his horse logging. He told me that he had worked for years as a U.S. Forest Service researcher, becoming an expert on lodgepole pine, before he retired and started horse logging six years ago.

We went back outside and walked past a trailer to a worn barn. Lotan fed hay to all the horses, then brought Belle and Duke forward with grain in his hand. He eased a bridle over each horse's head. "It takes a while to train these horses," he told me, "and it comes from dominating or controlling them without hurting them. You must get them to obey you without sweating and kicking and swearing. It's a bit like training a child to better carry himself." He told me that you need to communicate with these horses. You need, for example, to guide your hand all the way around their flanks when you walk around, so they know where you are at all times. You also need to try to assess what they are thinking, what is bothering them, so you can predict their behavior. "It all just comes from spending time with the animals," Lotan said.

He straightened a collar over Belle's neck, the big leather piece fitting like a good shoe. He hitched up Duke, then tied them both up as he went to get the arch—a carriage with two wheels, a seat in front, and a tall half-circle of metal behind. Chains running from the top of this metal arch keep the logs partly elevated while they are dragged. He put the arms of the arch up through the hames, curved supports attached to each horse's collar. They were ready.

Later, I saw a crew skidding with horses. A logger encouraged the horses right with a "gee," then left with a "haw" and a directional pull on the reins. He brought the team and its load up the hill with encouragement and a quick snap of the reins, "You've got it, you've got it, right up this slope." The horses' breasts tightened, their heads bobbed to the rhythms of the terrain. They became clouds of breath and body heat.

Horse logging is often more expensive than conventional logging, and generally takes two to four times as long. A horse team only removes about 1 to 2 mbf of timber in a day. But horse logging is elegant — horses can leave a single drag line where skidders make furrows and moguls. And horse loggers are not under as much pressure as other loggers to meet equipment payments with trees. "I've been logging for twenty years," said Ward Kemmer of Lincoln, Montana. "I used to be up in Alaska doing helicopter logging and felling trees 8 feet around. I recently fell into horse logging, and it's about as good as you can get for small-scale aesthetic logging."

In 1995, about four years after the horse logger left, Bill and Vicki hired Bob Love. They had read Love's letters to the editor in the local newspaper. One letter I remember accused Plum Creek Timber Company of "punishing the land with mindless and short-sighted forestry practices." The letter was so strong it felt like it burned from the newspaper into my hands.

The Byrd-Rinicks invited Love to look at their land. He walked the property for a day and then wrote a management proposal and contract. The short paper included costs for thinning and chipping and read like a natural history essay:

> The existing forest is dominated by lodgepole stands, 60 to 65 years old, created by the Halfmoon Fire in 1929. Past logging has also created 20- to 30-year-old and 10- to 15-year-old stands. . . . Old larch stumps indicate that larch dominated the forest at one time but were removed by harvesting before the fire. . . .
>
> Even though the dense lodgepole stands seem to be barren and stagnant, I think they are serving as travel corridors and security areas for wildlife. This forest may be an important link for wildlife between Teakettle Mountain across the river and the Desert Mountain, Apgar and the Lake Five areas. . . .
>
> The management for this forest should maintain the original larch forest while also preserving, in places, the lodgepole community.

The Byrd-Rinicks requested that Love harvest unmerchantable timber, to improve the character of the forest, until his expenses matched their income from the commercial harvesting. This arrangement allowed the Byrd-Rinicks to improve the merchantability of their forest for the future but not be thrown into a higher tax bracket with income on top of their retirement earnings. When the Byrd-Rinicks filed their taxes, they subtracted the expenses of hiring Love from the capital gains income that Love reported from the merchantable logs (see Chapter 3 for more information about taxes).

Vicki and I walked into the forest to see what Love had done. "After he left this place," she said, "we told him that he's the forest steward and can come back whenever he wants, to do whatever he needs."

We looked at one area, a few hundred feet, containing harvested, thinned, and untouched tree stands, depending on predominant tree species, density, and age. "This is classic Bob Love," she said. "He'll just give and take with the property, leaving some trees and taking some." Love leaves trees in places where animals shelter and cuts heavier where animals feed, or where shade-intolerant western larch or ponderosa pine need to regenerate. Love will maintain a staggered canopy where wild mushrooms grow, leave big snags for nests where he finds pileated woodpeckers, and provide a mix of open feeding ground and closed shelter where deer and elk live.

We walked close to the river, and she told me Love didn't touch anything within about 100 feet on each side of the water. "The Streamside Management Zone law puts limitations on the amount of wood people can harvest in a 50-foot area along primary streams," Vicki said. "Bob decided to take that a few steps further."

TECHNICAL NOTE

What Are the Laws of the Land?

All the intermountain states have forestry regulations to prevent slip-shod logging and roughed-up land. Oregon was the first state in the region to regulate private forestry with its 1972 Oregon Forest Practices Act, now emulated in twenty-six other states. The law, rooted constitutionally in interstate commerce regulation, was passed with the recognition that "private forest lands make a vital contribution to the state by providing jobs, products, tax base, and other social and economic benefits, by helping to maintain forest tree species, soil, air and water resources, and by providing a habitat for wildlife and aquatic life."

The Oregon regulations require landowners to submit timber-harvesting plans to the Oregon Department of Forestry before they can begin logging. The regulations cover everything from road-building and harvest plans to reforestation, slash treatment, herbicide use, and protection of wetlands and streams. Washington and Idaho have similar regulations (see Appendix I for the addresses of the state departments of forestry, where you can obtain copies of the regulations).

Montana, however, has a mainly voluntary program. Private landowners in the "freeman" state of Montana can do whatever they wish, aside from the Streamside Management Zone (SMZ) law, which regulates logging around streams, and the Fire Hazard Reduction Act (FHRA), which requires disposal of large-diameter slash.

The SMZ is the area within 50 feet along each bank of primary streams (streams with fish populations, or streams that flow for six or more months of the year and drain into another stream, lake, or river) or within 100 feet of a stream on slopes greater than 35 percent. In this zone, at least 50 percent of the trees greater than 8 inches dbh or 10 trees per 100 feet, whichever is greater, must be retained.

FHRA, the fire reduction law, requires residual slash to be less than 1 foot deep after harvesting. Loggers must pay a percentage of the timber harvest into a state fund, to be held until the land is inspected. The inspection ensures that 90 percent of the total harvested area meets a photo series where flames would be less than 4 feet tall if the area burned under conditions of 87 degrees Fahrenheit, 17 percent humidity, and 12-mile-per-hour winds.

In addition to these regulations, Montana has an extensive educational program for loggers and landowners that promotes best management practices (BMPs). BMPs reduce erosion from logging roads and harvests. These practices include, for example, the installation of drain dips in the road to carry water from the inside edge of the road to the outside, and the placement of rocks or timber slash to slow draining water. BMPs are purely voluntary, but if forestry practices discharge significant sediment into waterways, the landowner may be held liable for violations of the Clean Water Act, punishable by fines from the Environmental Protection Agency.

Recently, the bull trout, native to the Intermountain region, was listed under the Endangered Species Act, so regulations on streamside logging may tighten.

"The original guidelines of 50 feet on each side of the stream were probably based on the distance for trees to provide shade and control the temperature of the streamwater," said Chris Frissell, associate professor at the Flathead Lake Biological Station and a member of a committee evaluating the status of bull trout in Montana. "We have found that this is probably not enough to maintain our fish populations. We may need 100 feet or more in total to provide woody debris for pools, leaves and needles for nutrients in the stream, insect habitat for the feeding fish, and vegetation traps for reducing erosion."

"I agree that we need some regulation to protect our common resources such as air, water, wildlife, and trees," said Terry Lamers, state issues coordinator for the Oregon Forest Owners Association. "But if some of our landowners couldn't harvest on 200 or 300 feet around their streams, they basically couldn't do anything with their property. Regulations will tie their hands."

Washington state has tried to meet both scientific and social concerns. The state has extensive set-asides to protect salmon habitat— 175 feet of reduced cutting on each side of a stream, including 50 feet of no cutting at all. The state has also set up a Small Forest Landowner Office through the Department of Natural Resources, which helps landowners develop site-specific plans to protect endangered species while still meeting many of the landowners' management goals. In some cases, for example, the state buys 50-year riparian easements where they compensate landowners at 50 percent of the stumpage value for leaving stream buffers.

Maine has considered adopting the country's strictest forestry regulations, restricting clearcutting on private land without compensation to landowners. "Do you favor," Referendum #2 read in the fall of 2000, "requiring landowners to obtain a permit for all clearcuts and to define their cutting according to the Tree Growth Tax?"

The legislation failed, but states around the country are proposing similar legislation, such as California's Heritage Tree Preservation Act, which would have outlawed cutting trees of a certain age and size. The proposed rule in Maine had two major components:

- Landowners would not be able to cut more than their forest's annual growth rate. "The yearly allowable cut levels may not be greater than the average annual growth during the past 10 years."

- Landowners also would not be able to clearcut more than five acres without getting approval from a panel of scientists appointed by the governor.

Forestry is still Maine's primary industry, and tension over the referendum was tight, like a fist gripped behind every conversation. "This proposition would cause bad forestry," said Abby Holman of Maine's Forest Products Council, a group opposed to the referendum. A branch of the council set up to fight Referendum #2 included industry groups (such as International Paper, Fort James, Georgia-Pacific, and J. D. Irving), the Small Woodland Owners' Association of Maine, and the National Audubon Society.

Landowners should not plan their harvests from past growth rates any more than they should drive only with their rearview mirrors, the council said. Landowners need to be able to use computer projections that predict future growth, as well as using past growth rates. Landowners must be able to bank, so those who do not harvest every year can harvest a few years' worth of growth at once. Also, in older or diseased stands with stalled growth rates, harvesting is still needed to open up space and jump-start new growth.

Only 3.5 percent of the total harvest, or 18,754 acres, was clearcut in Maine in 1999, compared with 145,000 acres seven years ago, the Council notes. "We've already had extensive regulation over the past three years," said Holman. "We just finished a comprehensive review of the Maine Forest Practices Act, we have new rules for maintaining water quality, and we developed statewide benchmarks for sustainability. We're not against guidelines to improve forestry, but we're against rules that hold forestry back."

Already landowners need approval from the Maine Forest Service if they wish to clearcut more than 75 acres and it would be burdensome, weighty, for landowners and managers to get approval from a board, then go through a public review process, for small clearcuts. The referendum would hurt mom-and-pop NIPF owners across the state, the group argued.

"This referendum would apply only to those who are enrolled in the Tree Growth Tax Program. It's just asking people who get some benefit from the public to give something back," said Jonathan Carter, director of the Forest Ecology Network in Augusta, Maine, which promoted the legislation. The Tree Growth Tax allows landowners to pay a lower property tax rate if they keep their land forested. The tax was invoked to encourage "sustained yield" forestry and protect forests for "their unique economic and recreational resources."

Nonindustrial Private Forests (NIPFs) won't be affected by the new law, Carter said. Although large industrial owners subscribe to the Tree Growth Tax, only 20 percent of the NIPF owners in the state subscribe, as it saves landowners just a dollar or two per acre compared with agricultural property taxes. Further, according to a

A clearcut on industrial land in Maine

recent U.S. Forest Service assessment from 1982 through 1995, NIPF owners in the state already cut below the growth rate, while industrial owners cut twice as much as is growing back.

"Have you seen the Maine countryside?" Carter asked. I said that I had. A week earlier I had gone out to some industrial clearcuts with Bill Butler, a logger now retired into corduroys and an old farmhouse, who started logging the Maine forests in 1948. The forest looked like a dog with mange, tufts of coat coming out. "I can tell they've cut too hard because raspberries have started to come in here," Butler told me at one of the clearcuts we visited. "I can't help but look around and say that this could have been done better, that there are better ways to manage a forest."

To guard against violating regulations, you should contact the forestry office in your state well in advance of a timber harvest and meet with state foresters on the site to make sure that all pertinent regulations are being followed. Also, you should follow voluntary programs, such as BMPs, hire a reputable logger, contact the U.S. Fish and Wildlife Service regarding endangered species, and use sustainable forestry practices described earlier in the chapter.

Vicki and I continued our walk through Love's logged stand. She said she has referred some of her relatives and neighbors to Love, and he has met all their expectations. "There was only one odd case with my brother, Chris Byrd," she said. "He has about four acres of forestland and didn't like the aesthetics after Love's harvesting. He said there was just not enough sunlight in his forest, so he hired another logger to come in and sort of gut it out."

Vicki and I walked by a blowdown on the way back to the house. Love had told me about this place, where the wind had come in strong from the east, opposite the prevailing direction, after a wet summer. The surprise wind nailed flat dozens of Douglas-fir and lodgepole pine that Love had left around the harvested area. He gritted his teeth when he told this to me, angry that he had lost the trees. "I'm not perfect," he said. "Losing a few trees like this is a risk you take with selective logging."

Vicki also told me they had not made any money from this harvest, though they will profit from larger trees Love plans to remove in ten to twenty years. Trees take time to grow, she said, and forests must be considered ecologically rather than just in financial terms. Love's methods are best, she said, for landowners who already have solid income and wish to improve the merchantability of the land for their heirs or for future harvests. "Love's work was basically an investment for the future production of this forest," she said. "After the rough logging we had before, he put the forest back in order. He gave us redemption."

Douglas-fir
Pseudotsuga menziesii

Short evergreen needles, arranged in flexible rows. Bark has corky texture, colored reddish or blackish brown, with resin blisters on younger trees and deep furrows on older trees.

This tree, with a small subspecies growing in the Rocky Mountains and a larger subspecies growing in the Pacific Northwest, provides more lumber for our country than any other tree. Whether you buy plywood or two-by-fours, you are probably buying Douglas-fir.

Douglas-fir is primarily a timber tree, providing much of the wood we use, so it is appropriately named after human explorers. The tree was scientifically named for the Scottish physician and naturalist Archibald Menzies who first wrote about discovering the tree in 1791 while exploring the Pacific Northwest coast with George Vancouver. Its common name comes from David Douglas, a Scottish botanist who sent seed from the tree back to England in 1827.

This tree fits into its own family. It has needles in single rows like spruce, but the needles are soft; its cones are bracted like those of true fir, but fall from the tree before disintegrating. The bark on older trees is as thick as ponderosa pine and western larch, but is furrowed rather than plated.

In higher habitats, where Douglas-fir is endemic, it is one of the landowners' most dependable and merchantable species. But at lower elevations, shade-tolerant Douglas-fir has grown up into the overstory and prevented shade-intolerant larch and ponderosa pine from regenerating. Regular fires in the past killed most of the fire-susceptible Douglas-fir, but now the species dominates some ponderosa forests, making the whole forest more susceptible to root rot and crown fires.

LONG-TERM FOREST MANAGEMENT

*"We are going to leave this place
better than how we found it."*
—DAVID AND KATHRYN OWEN, CONDON, MONTANA

*David and Kathryn Owen on their
north-central Montana forestland.*

OWEN TREE FARM

Condon, Montana

ACREAGE
160 acres in western Montana; Section 17, T20N, R20 W, 80 easterly acres purchased in 1967, 80 westerly acres purchased in 1971, owned jointly by David and Kathryn Owen

MANAGEMENT OBJECTIVE
"We wish to improve the species composition and health of our forest, while maintaining wildlife habitat and forest diversity for our children and grandchildren."

GEOGRAPHY
4,000 feet elevation, west-southwest-facing aspect, loamy-clay soils

CLIMATE
60-day frost-free growing season
30 inches annual precipitation

STANDING TIMBER VOLUME
Standing volume in 1998: 480 mbf (3 mbf per acre total, 1 mbf sawlogs greater than 10 inches dbh), Trees per acre: 800-1,000

STANDING TIMBER SPECIES

Eastern 80-Acre Stand
Lodgepole pine: 95 percent of total stems
Western larch, Douglas-fir, and others: 5 percent
Average tree age: 110 years
Average tree size: 6 to 8 inches dbh
Average growth rate: ⅗ of an inch per year
Number of canopy stories: 1
Initial basal area: 224 square feet per acre, reduced to 160
square feet per acre with group selection harvests

Western 80-Acre Stand
Lodgepole pine: 80 percent of total stems
Engelmann spruce: 10 percent
Douglas-fir: 5 percent
Grand fir: 2 percent
Subalpine fir: 2 percent
Western larch: 1 percent
Average tree age: 30 years
Average tree size: 6 inches dbh
Average growth rate: 1 inch per year
Number of canopy stories: 3
Initial basal area: 120 square feet per acre, reduced to 100
square feet per acre with precommercial thinning

UNDERSTORY VEGETATION DIVERSITY

Primarily pinegrass-duff interspersed with yarrow, pearly everlasting, pussytoes, arnica, juniper, dogwood, clintonia, bunchberry, serviceberry, kinnikinnick, and Oregon grape

WILDLIFE DIVERSITY

Mammal sightings: black bear, grizzly bear, mountain lion, coyote, elk, moose, white-tailed deer, badger, marten, weasel, chipmunk, ground squirrel, red-backed vole, deer mouse, shrew, beaver, mule deer, muskrat

Bird sightings: ruffed grouse, pileated woodpecker, downy woodpecker, hairy woodpecker, great horned owl, barred owl, gray jay, Stellar's jay, common raven, wren, cedar waxwing, hermit thrush, rufous hummingbird, black-chinned hummingbird, flicker, mountain chickadee, robin, nuthatch, junco

HARVESTING HISTORY

Eastern 80-Acre Stand

Pre-1990: No disturbance since a fire in 1890

1990-present: commercial ¼-acre group selection harvests combined with replanting to restore ponderosa pine, western larch, and white pine to historic populations; harvested 230,000 bf total (post and pole size); 23,000 bf per year; 288 bf per acre per year average (10 percent of standing volume)

Western 80-Acre Stand

1964: High-grade harvest in 1964, approximately 150 mbf removed

1994-Present: precommercial thinning to increase growth rate of merchantable fir and spruce

ECONOMIC RETURNS

Average gross income from group selection harvests on eastern 80 acres: $3,500 per year total; $44 per acre per year; $152 per mbf per year. Costs and taxes are estimated as 20 percent of gross income.

THE OWENS' FORESTRY

David Owen, a Forest Service ranger, and Kathryn, a high school biology teacher, recently retired to their 160 acres of mixed conifer forestland in the Swan Valley of Montana that they purchased in the 1960s. The Owens have predominantly a lodgepole pine forest mixed with Englemann spruce, Douglas-fir, subalpine fir, grand fir and western larch.

- The Owens manage their land in different ways according to different forest types. On their western, mixed conifer stand that was harvested in the 1960s, they thin with a 10-foot spacing similar to Goebel's single tree selection, but not removing as many merchantable trees or providing as much space for regeneration. On their eastern lodgepole stand that grew up from a fire in the early 1900s, they harvest with ¼ acre selection and plant a variety of tree species to replace the decedent lodgepole. They select tree species to plant based on historical records for the area, and use seed from trees on their property.

- The Owens also think beyond their lifetimes and their small acreage. After an adjacent parcel was clearcut and subdivided, the Owens established a conservation easement on the richest half of their property.

The first principle is to give thanks for the environment
and that's called "care."
The second principle is to use the environment wisely
and that's called "stewardship."
The third principle is to trade or barter any excess
and that's called "economics."

—Saying from Kwakiutl Indians of British Columbia

L OOK AT WESTERN MONTANA on a topographic map and you can see it like an
unmade bedspread with smooth valleys, then complicated Rocky Mountain gatherings. At the lower elevations lie rough lands of sagebrush and scrub
oak, occasionally flamed with lupine and Indian paintbrush. Higher up from
these oatmeal hills stand the evergreens and, around streams and rivers, a
flush of cottonwood or serviceberry. Arêtes stand above it all.

At first glance, the Swan River Valley looks no different than when I traveled through here as a young boy with my parents. But now, after two decades
of subdivision, parts of the countryside have so many roads and houses that
they look like suburbs without a town.

I visited David and Kathryn Owen in the Swan Valley last year to find out
how they managed their forest estate. The Owens recently decided to write a
conservation easement with the Montana Land Reliance to reduce estate taxes
and protect their family land from development. They were voted 1997 Montana Tree Farmers of the Year by the American Tree Farm System.

I drove up through a maze of roads into a full forest and stopped at a gate.
The Owens had asked me not to drive up the muddy road and rip it up so I got
out of my car and walked (a scrawled message hung on the gate: "Route closed
for spring breakup"). On the walk, I heard the faraway sound of wind among
the treetops, then later felt the sudden rush around my body. It was a westerly
wind just beneath the mountains, a harbinger of rain.

David and Kathryn, both in their sixties, were outside when I walked up to
their cabin. "How was your drive up from the highway?" David asked, after I
shook his hand. I remembered the road, full of twisting turns, with my car
jerking back and forth on dirt and gravel.

"Okay," I said and smiled.

"In the winter our road is closed and we drag everything in on a plastic
sled—groceries, gas for the saw, the works," he said. The isolation of the property has allowed marten and mountain lion to come in. David has seen signs
of the animals—tufts of fur, fading tracks, flashing eyes.

"I saw a mountain lion once," Kathryn told me, and pointed out the window to the trail where she saw it. "So I snuck up a little closer and snapped a
picture of it. But after the camera flashed, the lion started coming toward me."
She paused, reliving the fright of the experience.

"I was waving my arms and shouting stupid things at him but he just kept moving toward me," she said. "I knew I could at least get to the outhouse, so I backed up the trail a bit, still facing the lion, then turned around and ran up the trail like you know what. I stayed in the outhouse for a terrible long time. David finally noticed me gone and looked out from the house. I caught his attention. He got a pan and spoons and started making noise to scare off the animal."

We looked out at the forest, as if we could see the impression of the mountain lion from years ago, then went inside the cabin and sat down to grilled cheese sandwiches and tomato soup. The warmth and food made us talk and smile.

"I've worked for the Forest Service as a ranger for 35 years and I've seen a lot of changes during that time," David began. "I don't like what I see around us anymore. I don't like these old-fashioned, grease-and-sawdust-oriented projects to remove tons of timber, or these plans for golf courses and ski hills. I want something in between, something stewardship oriented where we can get some of the wood we need while still retaining our forests. That's why we bought this place, to see if we could create what we envisioned."

I nodded and looked around the cabin. It was what I imagine a homesteader's cabin to be, with exposed timbers and chinking and piles of miscellany — old steel traps, snowshoes, a kerosene lantern, animal skins, a nest of blankets. During spring and summer, when the Owens live here, they store food in an ice chest, haul in jugs of water, and cook food over a wood-burning stove.

"So who does most of the work in the forest?" I asked, warm from the soup.

Inside the Owens' cabin

"Both of us," David and Kathryn said together. Kathryn has learned to run the chainsaw and, over the last decade, the two of them have precommercially thinned 35 acres of mixed conifer forest and cut 20 acres of lodgepole in quarter-acre group selection harvests.

"We manage this land carefully for wildlife," David said, proudly. "And in this forest type, unlike the lower-elevation ponderosa forests, Douglas-fir and grand fir aren't fire tinder—they are bedroom trees for animals."

In their western mixed species forest, the Owens keep the land lush, thinning down from 140 square feet per acre to 100 square feet per acre basal area, which provides continuous forest cover for animals such as lynx, marten, hairy woodpecker, and pileated woodpecker. The Owens keep approximately 10 feet between trees, but with shade-intolerant lodgepole pine, 10 feet is too tight for regeneration, so the Owens are primarily boosting the growth of crop trees. In the winter, typically an off-season for harvesting, the Owens prune some of the dominant Douglas-fir, grand fir, and Engelmann spruce that they have chosen for crop trees. They leave the pruned branches on the ground and black lichen, which covers the branches like a rich sauce, nourishes wintering deer and elk.

 TECHNICAL NOTE

How Do I Manage My Land for Wildlife?

"Landowners need to get to know their land and the animals that live there," said wildlife biologist Charlie Craighead from Jackson, Wyoming. "Get a few field guides and spend some time walking the property through the seasons. Find a professional who can help you, either a scientist at the state wildlife office or a private consultant (you can get specific contacts from state forestry offices or extension services, which are listed in Appendix I). Next, try to minimize your impact on the animals—you might concentrate your buildings and roads in one section of the property, for example. Finally, try to develop a broad range of forest habitat for animals. For example, a forest with trees of multiple ages will provide both early successional habitat for elk and late successional habitat for marten."

Fences can alternately encourage and discourage animals on your forestland. Barbed wire fences may need to be modified to allow for migration of deer and elk. Stays should be placed every 5½ feet and the top two wires strung 1 foot apart so the wires don't twist into the animals' feet when they try to jump over. The bottom wire should be placed 10 inches off the ground for animals to crawl under. Around gardens, lawns, and flower beds, hardy fences will need to be built for excluding wildlife. These fences should be 7 feet tall with at least two strands of wire above and two courses of woven wire below, with the woven wire or mesh buried at least a foot deep to deter burrowing gophers.

Animals are typically more treat than trouble, but if they do prove bothersome, you should first look for ways you may be causing the problem. Are the compost, garbage, and pet food all securely sealed off? Are your pets restrained? Have you blocked off passageways into the house or buildings where raccoons, skunks, squirrels, and bats might enter? If animals continue to be a problem, you may need to poison or trap them (contact state wildlife officials for local policies).

If you want to bring birds around your house—from mountain blue-birds spending the summer to cedar waxwings coming down from the north for the winter—you can set up birdfeeders and stock them with a seed mixture available at your local hardware store. If you bring birds in close to your house with a feeder, you should put stickers or cutout silhouettes on your windows to break up the reflection. Be aware that birdseed, as well as hummingbird feeders filled with sugar water, can attract bears. Also restrain your dogs and cats, particularly during the spring nesting season.

Another step to maintaining bird habitat is preserving interior forest. Neotropical migrants, such as Swainson's thrush, Townsend's warbler, yellow-bellied flycatcher, and barn swallows, fly to the Inter-mountain West in the summer and to the tropics in the winter. They require solid blocks of forestland ¼ acre or larger without roads or heavy cutting to protect against predation by domestic cats, raccoons, and skunks, and to protect against the brown-headed cowbird—a stealthy predatory bird that camouflages its eggs in the nests of other bird species.

Finally, you can landscape with native plants that birds and other animals enjoy, such as:

- Mountain ash, chokecherry, and serviceberry for fruit eaters

- Bitterbrush and rabbitbrush for seed eaters

- Pine, fir, and blue spruce to provide high cover

- Snowbrush, juniper, fescue, wheatgrass, sunflowers, and wild-flowers to provide groundcover

In the eastern lodgepole-dominated forest, the Owens cut ¼-acre openings. They keep ¼ to ½ acres of pure forest between the cut areas as screening and cover. Also, David and Kathryn leave strips 100 feet wide around the travel corridors—usually swales where they see tracks of animals moving over snow cover—so their harvests don't expose traveling animals.

The combination of forest types—mixed conifer and lodgepole—and harvesting treatments—thinning and group selection—give the Owens a variety of wildlife, from bounding deer to wobbling black bear. Game animals such as elk can bed in the mixed conifer stand, then go out and feed on grass in openings in the lodgepole stand.

"Let's go outside," David said, pushing himself up and away from the table. "I've got some cleaning up to do in the woods over here so maybe you can help me and then go see Kathryn later." I followed him as he sauntered through the forest. We walked first through a verdant forest of Douglas-fir/subalpine fir/Engelmann spruce. Then we walked past rows of lodgepole pine, linear as soldiers.

"This is one of the lessons I learned from owning land here," David said, stopping by an eccentric patch of Scotch pine. "We saw some people down valley growing Scotch pine as Christmas trees and really making a killing from them, so we planted 3,000 Scotch pine for a quick return for our children's education. Boy, what a mistake! Some of the Scotch pine picked up gall rust from the lodgepole. Some of the others would grow up tall and fast, just like we wanted, but they couldn't root into our clay soils and would just topple over. Knapweed has been a problem since an old portable mill was left in this area and tearing up the ground to make a bedding surface just made that worse. After the Scotch pine experience, we learned not to force things so much. We now try to work with the species we have."

Dave wore brown trousers, boots, a button-down shirt, suspenders, and a cap that read "Big John's" in yellow letters. We walked into the dry lodgepole forest together.

We walked up to a Ford truck with boards along the sides of the flatbed. It had a V-8—one of the old 1960s engines that could start out in third gear. A few cut trees lay like sticks through the area. Owen started cutting some of the trees into firewood, and I lifted the logs up and put them into the back of the truck. I soon fell into a pattern. I'd pick up the logs, then put them in the back of the truck, then go back. Back and forth, back and forth.

Near where Owen was working stood a 1950s-vintage tractor he uses for skidding and a skidding pan made from the battered hood of a car. He sometimes places the pan under the butts of the logs to make them slide more easily to reduce soil disturbance. He's attentive in the woods; he will pick up a mis-shapen tree limb, something most people would look right over, and take it home to carve a toy for one of his grandchildren.

"So, what are you trying to do here?" I asked.

"We've set a few goals," Owen said, opening his arms as if to encompass the whole land. "We wish to: maintain every tree species we originally found in the forest (even if it involves some planting); preserve the health of the timber (keep the dead and dying trees below 1 percent of the standing volume); improve timber growth and merchantability (boost growth rates up to 1 inch per year); and do our own work."

The stocking on this 160-acre lodgepole forest is 480 mbf total—2,000 bf per acre of posts and poles (less than 10 inches dbh) and 1,000 bf per acre of sawlogs (less than 16 inches dbh). This translates to 3 mbf per acre compared to 14 mbf per acre total in Goebel's fir forest. The Owens remove about 23 mbf per year from the eastern 80 acres for commercial sale—dramatically less than the 72 mbf Goebel removes from his 160 acres, but still about 10 percent of the total standing timber volume, compared to Goebel's removal of only 3 percent of the standing volume.

The Owens plant on the eastern 80 acres dominated by lodgepole pine. David mentioned a study by U.S. Forest Service researcher Jack Losensky that reconstructed the species in this area from old milling records in the early 1900s. Western larch, ponderosa pine, and western white pine, now less than 10 percent of the Owens' species mix, used to constitute 25 percent of the forest, before fire suppression and heavy logging.

"Kathryn and I understand that we need to be careful about these figures," David said. "The mill records don't adequately account for populations of less merchantable species. Also, this particular stand developed from a fire 110 years ago, so it's very much a natural feature of the disturbance regime in this landscape. But we can't have a big fire here anymore with all these houses around, and these lodgepole are prime for bark beetles, so we decided to harvest and bring in a more diverse species mix for the future. We don't like the influence of clearcutting on the soil and wildlife, nor can we use single-tree selection in lodgepole without a bunch of the trees falling down from wind, so we decided to do group selection."

The Owens plant throughout the group selection cuts, putting the seedlings between patches of regenerating lodgepole, and occasional spruce and fir. For the trees the Owens plant (usually western larch or ponderosa pine), they collect the seed from cones of tree species in the forest, so the seeds are adapted to the microconditions of the site, then send the seed to a commercial nursery to get started. After a year, the Owens plant the seedlings in appropriate areas—such as dry and open places for ponderosa pine. "We don't have a magical figure for how many trees per acre we plant," Kathryn explained. "We just plant what we feel comfortable with, which generally ends up being 10 to 20 feet between each tree." The Owens often cover the seedlings with protective yellow netting and stake the netting into the ground so the deer won't pick it off.

The Reforestation Tax Provision of the federal income tax gives the Owens pecuniary advantages for replanting. They can deduct 10 percent of their refor-estation costs immediately from their pretax earnings for a tax credit up to $10,000.

We walked into the western lodgepole/spruce/fir stand, the other 80 acres. "You should have seen this before we started harvesting here," David said. "It was harvested about 30 years ago and left as a snaggly mess." Here, in the shade-tolerant fir forests, the Owens thin to a 10-foot spacing, removing un-merchantable trees to make room for their stronger neighbors. Mostly the Owens remove premerchantable lodgepole from this stand, which they use for fence posts and firewood on their property. The Owens leave Engelmann spruce, Douglas-fir, and grand fir for future crop trees. They are accelerating a natural succession process.

I looked at David. "So over half of the trees on the property are less than 8 inches diameter," I said. "How do you make a dollar selling these little trees?"

"We might make $90 a day dipsy-doodling around here, after taxes and operating expenses, which is a little extra income for retirement," he said. "Truth

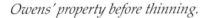

Owens' property before thinning.   *Owens' property after thinning.*

TECHNICAL NOTE

Economics

What Does Forest Management Cost?

Bud Moore, in his 80s, is one of the oldest landowners I met in Montana. When I visited him in the fall of 1999, he had burnished Redwings on his feet and cotton duck overalls covering his skinny frame. Twenty-five years ago, he bought these 230 acres of land, just a few miles from the Owens' place, and dug a foundation for his house with a shovel he called the "optimist." Before purchasing the land, he worked for the Forest Service and wrote a book about his career.

"Forestry has to involve economics, because most people can't afford to just play around in the woods," Moore said on our walk. "You need to make your licks count and make everything you can out of the situation.

"Montana has some bragging country with lots of timber bounty," he continued, "but you need to make sure you've got your objectives right so you don't end up with sour stuff. In the crowded corners of our woodlot, for example, my wife, Janet, and I took just a few posts and poles. In level spots with good soil where trees would come back, we took bigger sawlogs. We found out, like everyone needs to, how to set management objectives that fit the characteristics of our woodlot—how to make sustenance from the land."

Moore has developed a simple accounting sheet for his logging similar to that below. Such a sheet is useful if you do your own logging so you can identify areas to reduce costs, or if you hire someone else to log so you can easily compare bids. It is also important to keep records like this—along with receipts and mill tally sheets—for deducting expenses on federal income taxes.

Costs	Felling, Delimbing, & Skidding ($/mbf)	Hauling (if available)	Milling (if available)	Slash Disposal	Admin. (Plan Time)	Equip. (Rental & Deprec.)	Timber Stand Improvement (Pre-commercial Thinning, Prescribed Burning)

Income	Peelers ($/mbf & mbf)	Sawlogs	Posts & Pole	Firewood	Other (Consult, Equip. Rental, Nontimber Products, Hunt/Fish Permits)

Most of the landowners I talked with reported logging incomes and expenses similar to those listed below. The landowners cautioned, however, that there are ancillary expenses of landownership on top of these logging costs, including (for a 160-acre parcel at 2001 prices):

• Title searches by an attorney ($350 to $550)

• Boundary surveys by a forestry consultant ($500 to $1,500)

• Forest cruise by a consultant ($1,000 to $3,000 for a forest inventory and $1,000 to $2,000 for a management plan)

• Fence construction ($15 to $20 per rod (16.6 feet) for 4-strand barbed wire fence)

• Road building ($3,500 to $25,000 for the 160 acres, depending on the amount and quality of road, type of terrain, and durability of soil)

	INCOME	EXPENSES
Gross income from sawtimber-sized trees (10-16 inch dbh)	$300-600/mbf	
Costs of felling, delimbing, bucking, and skidding		$70-120/mbf (mechanized ground skidding) $120-220/mbf (horse, cable, or helicopter skidding)
Costs of precommercial thinning, prescribed burning, planting and other timber stand improvement		$50-500/acre
Costs of hauling sawtimber from deck to mill		$40-60/load/100 mi. ($5-15/mbf/100 mi.)
Net income	$0-525/mbf typically $100-300/mbf	

Real estate prices, meanwhile, range from $500 to over $20,000 per acre for undeveloped forest land, depending on whether the land is close to cities or resort towns, and whether it has particular scenic or recreational value.

What Does Sustainable Forestry Cost?

Admittedly, uneven-aged forestry practices provide smaller net returns than even-aged practices. Fewer and smaller trees are removed, and felling and skidding must be modified to reduce damage to surrounding trees. Some of the benefits of uneven-aged forestry, such as clean air and water, can't be pocketed. Other benefits––such as improved populations of wild animals, herbs, and mushrooms—can't easily be marketed to others. But the additional costs of sustainable forestry can be made up through:

- Boosted revenue from harvests timed with high annual timber prices

- Increases in the size and quality of the standing timber as it grows, joined with increasing timber prices (this totaled 4 percent annually, above the inflation rate on Goebel's property)

- Improved scenic and recreational value (clearcutting lowers property value by 30 percent or more, because of the loss of timber and aesthetics, while selective harvesting can improve land value by "cleaning up" dense timber; hunting or fishing permits might also be sold to generate additional income on selectively harvested land)

How Can My Forest Taxes Be Reduced?

PROPERTY TAXES

Counties go through a simple calculation to determine property taxes. The county adds the assessed value of all private property into a "grand list." The grand list is then divided into the total expenses for the county (education accounts for 60 to 80 percent of these costs) to determine a "mil" rate of tax dollars per thousand dollars of assessed value.

How can you reduce property taxes? Request that your forestland be assessed at its current use. On Bud Moore's Montana property in

1999, the forestland was taxed at a special agricultural rate of $1.24 per acre, while the house site was taxed at $124 per acre (100 times greater).

Most states have acreage or productivity requirements for landowners (such as owning 15 or more acres of contiguous forestland) to qualify for the forestland or agricultural rate. Some states also require management plans written by a professional forester. There are states that even allow yield taxes instead of property taxes so forest owners only have to pay taxes after a timber sale. Contact your state forester or extension service for more information.

INCOME TAXES

If you plan carefully, you can protect your timber income from high federal taxes. Timber returns (if they do not constitute a large share of your income) can be claimed as capital gains and reported on Schedule D like gains on stocks and bonds, rather than being reported as business income. The advantages are:

- Income from capital gains is taxed at a maximum rate of 20 percent, rather than almost 40 percent for other income (year 2001 rates).

- Capital losses from any source, such as losses from stocks, can offset capital gains, such as income from a timber sale, and reduce tax liability.

In addition to reporting income as capital gains, you should deduct costs "ordinary and necessary" to growing timber, including:

- Costs of owning land (property taxes, interest expenses, insurance)

- Costs of managing timber (equipment depreciation, forestry consulting, boundary maintenance, precommercial thinning)

- Original cost of the timber (this is called "retroactive allocation of basis" where present earnings are subtracted from the value of timber when the land was purchased)

Contact an accountant or attorney for more information on taxes, as individual situations vary.

is, we do a lot on a shoestring and we sweat a lot. It would take 400 acres of uncut forest to make a real living from lodgepole." (Maybe half that acreage would be needed for a lush fir forest like Goebel's.) David estimated that each of his 6- to 8-inch dbh lodgepole pine, about 75 board feet each, brings in $30 net income.

"I figure that what we're mostly doing is trying to get the place in shape for our children, so they'll have something good to work with, more than dying lodgepole in the eastern half and scrap from a poor timber harvest in the western half. We're paying dues now with hopes of a better forest for our children and grandchildren."

I met with Kathryn Owen after seeing Dave. She is a gentle woman, a grandmother who taught high school biology for several years in a small school west of Missoula. We walked together through a wet section of the forest, where we saw mushrooms of all shapes and colors, and ghostly Indian pipes — plants that grow in clumps and live saprophytically from decaying material on the forest floor. The plants look like vegetal jellyfish, clear and partly gelatinous.

"This place is full of interesting plants and animals," she said. "I've been studying these mushrooms lately." Her words came light and quick, full of energy.

We saw a *Boletus edulis,* or porcino mushroom. It had pinkish gills turning brown and felt slippery and firm. She picked it and put it in a wicker basket.

"Last year," she said, "I bought a microscope so I could see the spores and get down to closer identification. They're still a mystery sometimes." I confidently ripped off a piece of *Boletus* and ate it.

"I'll take this back with some others and identify them all to make sure we got them right," she said, winking at me as I chewed. We were standing beneath Kathryn's favorite hill, a place where she can get on top and see sweeping sections of the valley, from the Mission Mountains in the west to the Swan Range in the east.

Kathryn and I met David back at the cabin. The inside of the cabin smelled of wood and heat. I moved over to the kitchen stove and warmed my hands over the fire. I felt the tingle, then quick rush of blood back into my fingers.

"It's cold out there," David said. "This winter'll probably be cold as bones."

"Do you want some coffee?" Kathryn asked, putting a blue steel pot on the stove.

"Sure," I said. I sat down at the wooden table with David. We sat quietly for a while, listening to the stove pinging with heat and letting our skin settle to the temperature.

"I'm deeply concerned about what's going on in this valley," David said. "On the 1,000 acres north of us, they cut the middle right out of the land; they

■ MUSHROOM MADNESS

I met Larry Evans, president of the Montana Mycological Society, at the organization's annual meeting in Plains, Montana. Evans is a tall and thin man, with glasses and dark, curly hair in a ponytail. He wore baggy pants and a cotton T-shirt at the meeting and moved around so constantly, I had to jog around to catch up to him.

Evans and I talked at one of the mushroom display tables outside the hotel. He identified mushrooms on the table with names as impossible as their shapes and colors. We touched *Marasmius oreades,* the edible "fairy ring," and *Amanita bisporigera,* a pale-white mushroom with an annulus ring like a skirt half-way up the stem and a swollen base so poisonous it's nicknamed "Death Angel." I picked up the edible *Boletus edulis,* the same mushroom I had eaten at the Owens', and pressed my finger into its head to write my initials in the blue stain it produces. Evans grabbed a slimy yellow mushroom nicknamed the "Cowboy's Handkerchief." Another mushroom, with brown gelatinous folds, is called "Monkey Ears."

We both stopped at the end of the table to look at chanterelle and morel—valuable, wrinkly mushrooms. "These are good mushrooms to start out collecting because they are relatively common and difficult to confuse with poisonous mushrooms," Evans said. "But be sure, your first time, to go with someone who knows mushrooms."

Mushrooms are fruiting bodies of soil fungi, including the mycorrhizae fungi important to tree growth. Evans suggests always leaving at least one-quarter of a mushroom population in the forest to maintain the fungal populations.

It is also important to develop habitat for the mushrooms. The morel *(Morchella esculenta)* and chanterelle *(Cantharellus cibarius)* require partial sunlight and woody debris to grow. They often grow best on the edge of open or burned areas. After the fires of 2000, Montana had a bumper crop.

 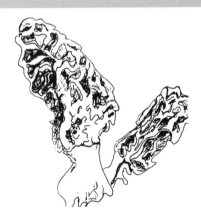

Chanterelle Mushroom (Cantharellus cibarius) Apricot-colored with wavy margins, rich in vitamins A, D, and carotene

Morel Mushroom (Morchella esculenta) Egg-shaped spongy or honeycomb head, swollen at base, also vitamin-rich

Other plants besides mushrooms can be valuable on your property. To strengthen your immune system and help prevent colds and flu, you could make tea from the flowers of yarrow *(Achillea millefolium)* or purple coneflower *(Echinacea angustifolia)* or from vitamin C-rich pine needles. Other non-timber forest products, besides mushrooms and medicinal herbs, are flowers, fruits, game animals, and specialty wood products like pine cones, wreaths, and landscape bark. Most landowners in the dry West cannot grow enough nontimber products to sell commercially. Still, landowners can use nontimber products themselves, or give them to friends, or sell them at local farmers' markets.

Evans and I cooked mushrooms later that day, slicing up chanterelles and putting them in a fry pan with butter. "Morel mushrooms are my favorite," he said as the mushrooms hissed in the pan. "I like them marinated in a teriyaki mixture of soy sauce, sesame, garlic, and orange juice." We ate the mushrooms just as they started to brown. The experience was less of flavor than of texture, a chewy softness.

stumped and diced the forest and turned it to pasture." He made a cutting motion with his hands on the table. "A developer from Florida came in and cut 3 to 4 million board feet of sawtimber and sold it. Then he piled enough lodgepole to run a local post and pole plant and just burned that. He pastured cattle on the empty land, then turned it over to subdividers."

David and I sipped on our coffee quietly. My breathing was fast and hot from his words. A goldfinch sang for a moment outside.

"Aren't you just doing the same thing this developer is, taking a piece of land and building your house on it and cutting some timber when you feel like it?" I asked.

"Look, I'm not against letting people do what they want with their own land," David said. "But there have to be some limitations, either ethical or legal. I mean it's one thing to cut a few trees for fencing poles or firewood, and it's another to cut and burn good timber. It's one thing to sell a couple house sites and keep some bigger acreages around it, and another to divide 1,000 acres of forestland into little lots."

David paused, then continued. "Last year, Kathryn and I went to the Montana Land Reliance to set up a conservation easement. We wanted to keep this land in place because we're concerned that everything else will be carved up around us. This easement is something we did for the land." David passed me a thick document. "We're doing this on 80 of our 160 acres, the most productive western acres. Then we'll probably try it on the other 80. We agree strongly with the intentions of an easement but we don't want to tie everything up right away. We're trying to balance the needs of this area ecologically with our family's potential economic needs."

The first thing I saw when I opened the document was an aerial photograph of the property—Section 17, Township 20 North, Range 16 West. The Owens' property looked dark against the developer's clearcut property directly to the north, which was barren yellow, like scrambled eggs.

"David and Kathryn talked to all of us before they set up the easement," said Colleen Owen, one of the grown children. "My younger brother will probably be the only one who would live out there as the rest of us have jobs and children elsewhere. But we all want to visit. We're glad that the land will stay in one piece. This land has been important to our whole family—we've all spent parts of our childhood growing up there."

"Where have you been?" David asked when I called again. I was comforted by his familiar voice and told him briefly about my travels.

We started talking about how other landowners might be encouraged to develop sustainable forestry practices. It could be forced on landowners by

Denuded forestland north of the Owens'

requiring them to submit timber management plans to state agencies, as is already required in Idaho, Washington, and Oregon. But administration is expensive and enforcement is difficult, particularly in a region like the Intermountain West so protective of private property rights. Other policy changes could reduce financial pressure on landowners so they don't need to cut so hard: reductions of estate taxes; replacement of property taxes with timber taxes; low-interest federal loans on standing timber that could be repaid slowly through selective harvests.

A number of other innovative programs are starting to promote sustainable forestry among landowners. (See Appendix I for more information.)

- In the Pacific Northwest, some landowners are able to lease out some of their property to harvesters who collect flowers, mushrooms, or medicinal plants through an organization called Rainkist.

- In the Southeast, some landowners, such as the private industrial company Anderson-Tully, have sold lucrative quail and deer hunting permits. Ducks Unlimited, Trout Unlimited, or the Rocky Mountain Elk Foundation may be able to help you with habitat improvement funds or in setting up a hunting and fishing permit program.

- Utilities and governments may soon be purchasing carbon emission credits from forest owners to keep their lands forested to reduce greenhouse gas emissions, as is already being done by the Pacific Forest Trust.

- Some landowners may also be able to sell their water rights to conservation groups wanting to protect instream flows, such as Water Watch.

- The Oregon Forest Resource Trust provides low-interest loans to land-owners for reforestation based on landowners' commitment to repay the funds when the reforested stands are harvested.

- The Trust for Public Land, a national nonprofit that helps conserve land for people, is working with large forest owners to help them develop conservation easements under the government's Forest Legacy program that helps defray easement expenses.

- The Nature Conservancy is trying to develop a Forest Bank program where landowners get a guaranteed annual return from timber harvests if they turn over their timber harvesting to the organization.

- Finally, a number of conservation organizations are buying forestland outright to protect it from development, then managing it with sustainable forestry to pay off the purchase. One such example is the Evergreen Forest Trust's recent purchase of the 104,000-acre Snoqualmie Tree Farm from Weyerhaeuser.

TECHNICAL NOTE

What Will Happen to My Property When I Die?

When you die, your estate—which includes land, houses, cars, stocks and bonds, insurance, and other assets—may be taxed at rates from 37 percent to 55 percent of the current market value. There is an exemption for the first $1 million (moving up to $3.5 million by 2009), minus depreciation, with nine months after death allowed for payment (though this may be extended under pressing circumstances). While only about 2 percent of estates are large enough to have estate taxes owed on them, real estate can quickly appreciate. Your assets could be worth far more at future prices than they are now.

"The basic problem is that without proper planning, a piece of land can trigger an estate tax so large that the land itself will have to be sold to pay the tax," says attorney Stephen J. Small of Boston, an expert on estate planning for landowners. "Rising real estate values have left many rural families land rich and cash poor."

The first step in estate management is to list all your assets, at current market value. If you are married you should divide all assets between you and your spouse so each of you has at least the amount of the estate-tax exemption (currently $1 million) in your name. Then, each of you should establish an A/B Trust, which upon death will pass assets up to the value of the estate-tax exemption to the children (beneficiary A) and transfer remaining assets to the surviving spouse (beneficiary B).

Next walk through your forestland and put in writing what you want done with the land. What kind of harvests, tree plantings, or recreational activities would you like to see? How, if at all, will the land provide for the family's financial needs? Which forest areas should be preserved?

Finally, consider methods for reducing estate taxes, which you should do in consultation with an attorney. You can protect your property from estate taxes and from development with family limited partnerships and conservation easements. These methods are discussed in detail below. (If you have over 1,000 acres of forestland, you could qualify for federal aid for estate planning through the Forest Legacy Program. Contact your state forester or extension service for more information.) Other options for protecting your financial assets include:

- Establishing a private foundation

- An irrevocable life insurance trust

- A charitable remainder trust, where you and your spouse specify an income you will get from the property until you die, then turn over the property to a nonprofit organization, such as a university, a church, the Nature Conservancy, or the Boy Scouts of America

- A restrictive covenant, which limits land use for a specific period of time, rather than forever as in the case of a conservation easement, but also gives fewer economic benefits

▶ Family Limited Partnership

With a family limited partnership or limited liability partnership, you or your surviving spouse will retain control of the forestland as "general partners," but you can divest much of your ownership in the forestland to "limited partners," your children. The limited partners will have no authority to make decisions, but will end up with the majority ownership relative to the general partners in order to lessen the tax liability upon inheritance of the remainder of the estate. Federal income tax law allows any individual to give assets as gifts tax free, up to $11,000 per recipient per year. If each spouse has a portion of the forestland partnership in his/her name, they can each give up to $11,000 of shares of the asset to each of their children every year. Further, the limitations on marketability of partnership shares (where the shares can't be sold or willed to anyone but another partner without consent of all of the partners) lowers or discounts the market value of the shares. A 30 percent reduction is considered common, which means that each parent can give approximately $14,000 worth of the partnership to each child, each year. A limited family partnership combined with a giving program can reduce estate tax liability and ensure that the land stays in the family without giving up managerial control.

▶ Conservation Easements

Conservation easements restrict development rights more than family limited partnerships do, but further limit tax liability and permanently protect the land. A conservation easement is established between a conservation organization, called the "grantee" (see the listing of organizations in Appendix I), and a landowner or "grantor." The easement primarily limits subdivision and new construction—conditions known as "covenants" that will run with the title of the land forever, even when it is sold or willed to other owners. A conservation organization will work with you to develop the covenants you desire, then come annually to the property to inspect it and ensure that the covenants are being followed. "Forest managers need to work with donee organizations to make sure that the landowner's interests and the specific conditions of the forest are considered in the contract," advises attorney Small. "The easement contract should have a set of measurable standards or guidelines for forest management, rather than a rigid management plan that may not cover all contingencies."

Landowners receive three economic benefits from establishing a conservation easement:

- Income tax deduction. The fair market value of a gift (the easement) to a qualified public charity (a conservation organization) can offset up to 30 percent of the landowner's adjusted gross income in any single year and be carried over for five years

- Reduction in the market value when the land is assessed for estate taxes

- Exclusion from estate tax of up to 40% of the market value of the property, with a $500,000 cap, depending on the location of the land

Conservation easements can be combined with family limited partnerships to maximize estate tax protection, keep the property in the family, and protect the property from development. (See Appendix III for more information on preparing an easement.)

	Family Limited Partnerships or Limited Liability Partnerships	Conservation Easements
Cost of set-up	Attorney fees from $1,500 to $5,000	Donations to donee organizations of $500 to $5,000 plus some attorney fees
Tax reduction	Each parent can give away $11,000 (up to $14,000 with market value reduction) in partnership assets each year to each child or other recipient, thus reducing estate taxes	Lowers the market value of the forestland when the land is assessed for estate taxes Another 40 percent of the market value rate, up to $500,000, may be excluded from estate taxes Offsets up to 30 percent of adjusted gross income from income taxes for up to five years
Restrictions on property	Can sell or will property freely between partners or to others if limited partners so vote or are bought out Can subdivide or build Can harvest timber, graze animals,and otherwise use the property	Can sell or will property freely, though easement covenants stay with title Can build on a limited basis and rarely subdivide Can harvest timber, graze animals and otherwise use the property, within certain restrictions laid out in the easement
Family ownership	Land typically stays in family	Land may or may not stay in family
Long-term conservation	Land is not protected, but restrictions (subject to modification by future partners) can be written into partnership	Land will be protected forever and be monitored by conservation organization

Landowners could also get their wood certified or join a forest landowner association. The Forest Stewardship Council (FSC), an international group with rigorous certification standards, works through SmartWood and Scientific Certification Systems to certify wood in the U.S. Although the certification process is expensive, FSC has a manager certification program where foresters can get certified, then pass that certification on to some of the properties they manage at a low cost. For now, certification is mainly a badge of honor, showing that landowners are doing right by their land, although it is an opportunity to find customers interested in "green" wood. Nine million acres have been certified in the U.S. by FSC—many more acres have been certified by an industry group, the Sustainable Forestry Initiative (SFI)—but with only 2 to 5 percent of lumber products being sold as certified today, it's hard for many landowners to get premium prices for their wood the way farmers do for organic food.

Over twenty-five landowner associations and cooperatives have developed across the country. Jim Birkemeier in Wisconsin started the country's first landowner co-op, the Sustainable Woods Cooperative, six years ago. He has practiced sustainable forestry on his property for 20 years—selecting the worst trees first, removing the big and straight trees when their growth stagnates, and removing all trees carefully, arthroscopically. The only way for him to offset the higher logging and forest management costs has been to do more of the processing himself and sell the timber as finished lumber. That's what he suggested to his friends and neighbors: let's band together, manage our forests in sustainable ways, and process and sell the wood ourselves. The co-op offers educational workshops and connects landowners to two FSC-certified forestry consultants in the state. The co-op not only helps landowners with management but also buys wood from them, which it processes on a 4-acre lot with a Wood-Mizer portable sawmill and a solar kiln. The wood is then often sent to a tongue-and-groove company to be made into siding and flooring. The co-op, which can sell its wood as certified and local, is often able to buy lower grade timber. "Being able to sell lower quality wood at the co-op can change a costly thinning into one that pays for itself," says Brad Hutnik of Clark Forestry. The co-op, which now has 150 members, sold $40,000 worth of wood products in 2001 and has already sold $140,000 in 2002.

A landowner organization called Vermont Family Forests also holds educational workshops on sustainable forestry, and helps groups members get FSC-certification. Instead of buying and processing timber, however, Vermont Family Forests simply links its inventory of wood from landowners' properties to customers who want certified wood. The organization recently sold tens of thousands of dollars' worth of lumber from their members' forests to Middlebury College for its new Bicentennial Hall. Co-ops like Sustainable Woods Cooperative and landowner associations like Vermont Family Forests could help many landowners with management and marketing, and interest in them is growing, marked by a provision for forest co-op funding that barely failed the final 2002 Congressional Farm Bill. Still, these organizations are new and experimental here, and have only become well developed in European countries and Canadian provinces.

"I'm not sure how sustainable forestry should be encouraged," David said. "Maybe it's more landowner education. Or maybe it's just an individual thing,

■ MULTIGENERATIONAL FORESTRY

Bill Potter, the landowner with beautiful ponderosa pine mentioned in Chapter 1, recently finished writing an innovative conservation easement with the Nature Conservancy. Potter has a body thinned from constant walks through the woods, and hair grayed from years of experience. He's growly and wise, and spits out epithets against industrial timber harvesting. "Now Plum Creek Timber Company is saying they only take the diseased trees," Potter said. "Well, most trees die of only one disease and that's the chain saw." Potter's land is bordered by Plum Creek land and he curses them every time he drives by.

"I've been working on this land for 30 years, and it's taken a lot of physical and mental work to put the forest together," Potter said. "I can tell you that when you're dealing with trees that take a century or two to grow, forest management takes more than one person's lifetime. That's why I put together this easement." Potter's innovative conservation easement involves the University of Montana Experimental Forest, the Nature Conservancy, and the E–L Ranch so that the management will be long-term and continuous, not jumping from one heir to another.

The easement reads: "At a minimum, the heirs, the Experiment Station, and the Nature Conservancy will meet annually on the first Thursday in September to jointly review the past year's activities, develop site-specific stand prescriptions for the following operating season and consider prescriptions proposed by one of the parties."

The maximum amount of live timber harvested, the easement states, must be less than the average annual growth as measured every 10 years in a series of permanent growth plots. (The rate so far in Potter's ponderosa pine forest is 180 bf per acre, or 1.2 percent, with a total stocking of 15,000 bf per acre). The amount of harvested material may be increased in cases of catastrophic wildfires, windstorms, or insect and disease epidemics. The easement calls for group selection of $1/10$ to $1/4$ acre in even-aged ponderosa pine forests every 20 to 30 years to convert them to uneven-aged stands; single-tree selection of uneven-aged ponderosa pine forests every 20 to 30 years to perpetuate the forest structure; and shelterwood harvesting (which harvests most of the trees over two to three cuttings in a 10-year period) of even-aged Douglas-fir forests to prevent the spread of spruce budworm.

Potter's easement also keeps constant the total acreage of old growth on the property (roughly defined as areas with substantial trees greater than 15 inches dbh and greater than 150 years old). Harvesting proceeds will be split—a portion will cover forest management costs, including consulting from the Nature Conservancy and the University of Montana, and the remainder will go to the heirs.

"Before Potter's easement we didn't have anything near this involvement in forest management in our easements. We basically left it all up to the landowners," said Bernie Hall of the Montana Nature Conservancy. "Now other easement holders are asking for our help in balancing resource use with protection."

and our children need to be brought up better to respect themselves and other living things." The Owens have tried to infuse a land ethic in their children by bringing the children out to the property to help pile brush, lay out research plots, mark trees, and fell and buck timber. The Owens have also taken the children on hikes through the Bob Marshall Wilderness, and Kathryn has taught them gardening, plant identification, and biology. Through this work and leisure, the parents say, the children began to understand and appreciate the land.

Epilogue

COMING OUT
OF THE WEST

"I can't cut more wood than I grow, just as I can't withdraw more money than I put in the bank."
— MEL AMES, ATKINSON, MAINE

Mel Ames in front of a woodpecker-poked snag on his Maine forestland.

"[C]onservation will ultimately boil down to rewarding
the private landowner who conserves the public interest."
—Conservationist Aldo Leopold in *A Sand County Almanac*

GO EAST PAST THE PINEY MOUNTAINS of Idaho and Montana. Go through the midwestern plains with squared-up roads and fields forced round by center-pivot irrigation. Go through the forests of Connecticut, with maples that turn acidic red and birches that turn ischemic yellow. Here lie the mixed softwood-hardwood forests of Maine, the largest contiguous stretch of forest-land east of the Mississippi.

Maine country, beyond bubbles of light around highway towns, is as empty and dark as a velvet stage curtain. Mountains are worn, as if they were wooden knobs rubbed down by oily palms; forests form painter's colors; the country-side, during mild winters, smells like a clove apple; and the lakes have ancient Indian names like Musquash, Nollesemic, Pemadumcook, Wesserunsett.

I drove up into the north woods from Bangor. Dipping in and out of hills, I saw mixed forests—tall white pine, bristly spruce, oak and maple large and straight enough for heirloom cabinets. The forests seemed to extend forever, without pause of plain or meadow. "Most striking in the Maine wilderness is the continuousness of the forest," wrote Henry David Thoreau in *The Maine Woods*. "Except for the few burnt lands, the narrow intervals on the rivers and the bare tops of mountains . . . the forest is uninterrupted." In a dream my first night in Maine, I saw house lot owners beating off invading trees with broom-sticks, and multinational paper companies (which made clearcuts as large as 32 square miles in the 1980s) shaving the forest flat with giant electric razors.

Mel and Betty Ames live in an octagonal house built with wood from their forest. The house stands like a punctuation mark in the middle of a hay field covered with snow. Up and down the road just a few rural houses stand, and, on a distant corner, sits Snow's Saw Shop, which sells gas and chain saws year-round, snow blowers in the winter and lawn mowers in the summer. A stripe of highway lies in front of the Ames' house, and behind stands a semicircle of forest.

I drove up to the house in the evening and parked by a gray Isuzu Trooper. Almost as soon as I got in the door, Mel's wife, Betty, offered me a place at the table. We had dinner with all the fixin's—green beans, mashed potatoes, roasted turnips, squash and carrots, bread—and afterward Betty offered me a slice of homemade pie—apple, pecan, or raspberry. "Feel free to pick one, a couple, or some of all three," she said. "Desserts are my cooking specialty." The pie tasted melty sweet.

I would come back to visit Mel and Betty one more time, and I returned as much to enjoy Betty's cooking as to learn from Mel's forest management. On my last visit, I had homemade carrot cake. "It's my mother's recipe and has just about everything in it," Betty said. "Coconut, rice crispies, raisins, nuts, you name it." Betty's abundant desserts almost matched a meal I ate at a landowner's house in Georgia where three cooks made a big spread of okra, collard greens, corn bread, macaroni and cheese, six different types of casseroles, and homemade ice cream and apple pie.

Betty cleared the table. "You missed a nice day today," she said. "Mel was out in his shirt sleeves. Usually we can't even see Mount Katahdin for most of the winter because it's covered in clouds like a wool sweater."

Mel Ames said he was born in a small town called Milo, near Atkinson where he now lives. When he was twenty, he bought this land. "I tried raisin' potatoes and dairy cows," he said. "We got ourselves a few prize cows but that was a lotta work, and raisin' potatoes wasn't a real moneymaker. So I tried managing the woodland—400 acres then, and about 500 today.

"It's worked real well for us," Ames continued. "I've been self-employed as a forestry consultant and forestry educator for most of my life, and managed over 2,000 acres of forestland all over this area for the state or for private landowners. I've brought up eight children on my income." The Ames' children range in age from 33 to 52, and live, still grouped together, in Maine and North Carolina. They teach school, nurse patients, manage restaurants, and run counseling centers. One of Mel's sons, Russ, is learning how Mel manages the forestland and will carry on the legacy.

So how do the children feel about Mel's work? How about the neighbors? "Everyone basically likes what I'm doing," Ames said. He had a shock of curly white hair on top of his head, and his voice cracked with enthusiasm. "I've shown that we can get timber and also the other benefits that go along with the forest. I've shown that it's possible to cut wood and still have oxygen, water, plants, wildlife!"

Later in the evening, one of Mel's friends, Charlie Fitzgerald, came by. Fitzgerald, a middle-aged New Englander, started a chain of stores called "Bowl & Board" which spread from New York City up to Martha's Vineyard. He uses odds 'n ends—stumps, lathe overruns, turnings, beaver-chewed bark—and transmogrifies them, like a magician, into candle boxes, birdhouses, back massagers, toy trucks, puzzles, and Christmas ornaments for his store. He picked up a piece of wood at Mel's table and turned it over and over in his hands, telling me how he would cut games, puzzles, and a spice rack from it.

The next day, I woke at seven and had a bowl of cereal and a cup of ginseng tea. Ames and I walked out the sliding glass door to an old GMC Sierra truck.

A hefty German shepherd, Kate, followed behind us as we drove out into the forest.

"I've worked the same land for 53 years now, so I have a good understanding of what works and what doesn't, what I did right and what I did wrong," Ames said. "Across the 520 acres of woodland, I've got some shade tolerant species that grow for over 200 years, like spruce and maple. I've also got some trees with partial shade tolerance that go for about 50 to 100 years, like fir and pine. Then, of course, there are those shrubs and trees that come in right after a harvest and need direct light, raspberries and birch. Gosh, I think I've got some of the most beautiful trees in the country.

"This is a really wonderful place," Ames continued, "I've got pine marten and moose, celandine and nightshade, jack-in-the-pulpit and dutchman's britches."

Part of the richness, Ames said, comes from the geography—nutrient-rich clays matched with a productive four-month growing season and 40 inches of precipitation per year. One part of the forest, for example, has glacial "kettles" where icebergs sat 10,000 years ago and drained out in "spouts," leaving lush depressions.

The diversity also comes from not overworking the forest. "There are only two reasons to harvest a tree," Ames said. "It's either 3D (dying, dead, or deformed), or it's around another tree of higher value that needs to be released."

Ames identifies certain trees as crop trees because of their straightness, lack of limbs, and the marketability of the species. He marks other trees as recruits to eventually replace the crop trees. Then he watches the trees in between to see how they interact with the crop trees. Do they nurse up some of the crop trees and reduce branching, or could they interfere with the crop trees' growth? Ames showed me one place in particular where he was cutting out some old hemlock, around 150 years old, and neighboring white pine was coming in, which not only is twice as valuable but grows twice as fast as the hemlock.

Ames continued driving the truck. Oil canisters, Pepsi cans, and oak and maple leaves shook in the flatbed. We stopped and Ames went to the back to get his Husqvarna chain saw. He pointed over to a maple with a frost crack, where rot might come in. "There's no sense in growin' a cracked tree like this," he said. "I've been lookin' at it and it's hardly growin' at all."

Ames' hands got busy with preparation: sharpening each tooth of the saw with a round file; adding red oil to lubricate the chain; tightening bolts on the front of the saw, just enough to let the chain run without bunching, not enough to clamp it up tight.

He first cut a small hemlock and leaned it down on one side of the maple to cushion the larger tree's fall. Then he brought the saw up to the maple. Wood fiber turned to dust. Except for a small hinge of broken wood, the trunk sepa-rated neatly from the stump like a piece of clay divided with a sharp wire.

The rings on the stump confirmed Ames's intuition that rot had infected the tree. "A few more years and this would've been nothing but firewood," he said.

Ames put the saw back onto the flatbed and we walked into the forest. The ash, oak, and maple stood bare, showing all the contortions and odd twists of their growth. The spruce and hemlock stood tall, covered with needles—furry clumps in the landscape of naked deciduous trees. I could tell the integrity of the forest by feel and sound as much as by sight: the slide of ferns and grass, the spongy give of rotted wood, the symphonic rattle of leaves and crunching wood.

Ames's forestry practices provide enough sunlight to encourage fresh di-ameter increases, and allow some new growth without encouraging a thick carpet of new trees that would need to be thinned. Ames said that he uses a combination of uneven-aged harvesting practices to develop his forest charac-ter. He employs commercial thinning on his 124 acres of hardwood to increase the growth of veneer trees (after he has kept them shaded enough to maintain a straight stem and prevent epicormic or stem branching); group-selection harvests on his 121 acres of softwood, where he removes up to ⅙ of an acre of trees for pulp; and single-tree selection on the mixed 263 acres of the property (where he keeps trees spaced 15 to 25 feet apart, a spacing that keeps the trees "growing good and juicy," as he put it). "I try to work with natural processes and succession," Ames said, "I take pioneer species that are going out . . . and help other species that are coming in."

Ames recognizes that in eastern forests, single-tree selection is often the best choice for beech/birch/maple forests, where sugar maple is favored and spruce/fir forests, where red spruce is favored. These harvests mimic natu-ral mortality from beech bark disease and spruce budworm.

In areas where other species need to regenerate, varied group selection might be used. Small openings of ⅒ of an acre allow regeneration of sugar maple, beech hemlock, and spruce; medium openings up to ⅔ of an acre allow new yellow birch, white ash, and red maple; large openings of up to 2 acres allow new paper birch, aspen, and pin cherry.

In addition to trying to balance the size of the cut to the composition of the forest, Ames must follow the forestry regulations for the Township of Atkinson, among the most stringent in the country. The regulations reduce the amount of timber that can be taken, although the main motivation for passing them was probably to keep timber in Maine—most of the harvesting here occurs on private land by companies who ship the logs off to Canadian mills for processing.

Ames and I got in the truck, the dog prying her way into the cab between my legs and the door so she wouldn't have to walk any farther. We drove a little ways, then stopped by a yellow cable skidder. A bumper sticker on the side of the cab read "Ban Clearcutting." The skidder, a John Deere 440, was only slightly larger than the skidder Leo Goebel uses on his property in Oregon. This skidder's size makes it more efficient than a tractor because it can haul more, Ames said. It's less expensive and needs less space to maneuver than big grapple skidders. Ames delimbs the trees in the forest and bucks them down to 25-foot sections before hauling them off. He also uses "bumper" trees, which are trees on the corners of skid trails that the wood can be knocked around to protect more merchantable wood behind it.

We got back into the truck and drove out to another area of the forest where we could get out and walk. Ames leaned one arm against a spruce. "These are the softwood flats of spruce, fir, and pine," he said.

I thought that these were his primary crop trees. After all, softwoods grow up straight and fast and regenerate easily compared to hardwoods. They can also grow on more acidic soils and require half as many nutrients as hardwoods. Ames said that he used to favor softwoods, particularly hemlock, for paper pulp, like many of the industrial landowners in the area. But hemlock is allelopathic, meaning that it produces chemicals in its roots that kill off other

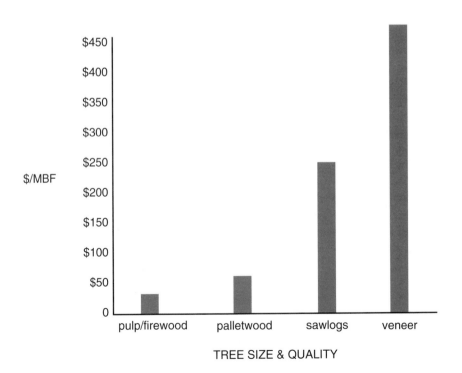

TREE SIZE & QUALITY

Trees of all species increase dramatically in value as they increase in size. This trend is particularly distinct for eastern hardwood species valued for furniture veneer, such as the hard maple shown here. — Information from Annual Stumpage Report, Maine Forest Service, 1998

tree species. Also, an insect pest that is spreading north from the mid-Atlantic states, the woolly adelgid, severely weakens and kills pure hemlock stands. Finally, maintaining a diversity of tree species in the East is a guard against acid rain, which leaches calcium and magnesium from the soil and reduces tree growth by damaging leaves and needles.

Now Ames is trying to restore the species diversity of his forest so he can grow hardwood veneer trees alongside the conifers. Ames said that he can make more from six 90-year-old hardwood veneer trees than he would make from an entire acre of 50-year-old pulp. He can make as much cutting two trees per day of furniture-veneer-quality hardwood as he would make cutting 100 trees per day of paper-pulp-quality softwood.

"Look, pulpwood that I would get from doing a series of group selection cuts is just a bread-and-butter operation, where I only make money on volume. I need to remove the pulpwood so I don't end up with a whole forest of the stuff, but I also need some of it because not every tree can be a veneer tree," Ames said.

"Veneer is the primary reason I grow big trees on the property," Ames continued. "If we have a 16-inch tree, that'll give us a return of about $238. An 18-inch tree will yield $297. So, at a third of an inch of growth per year, it'll take six years to get those 2 inches, which is a return of over 8% per year!" Of course, growing veneer involves "oodles of work," such as monitoring, thinning, pruning, and selective cutting, which take away a percentage of the returns.

In an average year on his 520 acres, Ames will cut 2 cords of pulp, 90 mbf of sawlogs, and 4 mbf of veneer, for a net income of $12,560 per year to add to retirement. Ames now generally harvests 98 mbf/yr, only 1.4% of the total stocking of 6500 mbf, but some years, when timber prices are high, he harvests up to 200 mbf/yr. Ames, like Goebel, has boosted his timber volume by cutting out junky trees, doubling his per-acre stocking over the last fifty years.

Woodlots should be managed profitably, but as long-term investments, Ames emphasized. We walked next to a "big 'un," a 200-year-old oak. "I could get $500 of timber from that one easy, but I'll let it die here," he said. "It's a nice tree for a snag." Ames pointed to a couple smaller trees nearby that he might take for timber.

Betty put out a selection of cold cereals and milk for me the next morning.

"Mel," she said, "it wasn't too many days ago when I'd fix you a peanut butter sandwich with a glass of milk at 5:30 every morning."

"That's right," Mel said, "It didn't hurt me none, either, kept me going all day. That was back before we had that skidder, back when I felled wood with bucksaws and twitched it with horses."

As I finished my cereal, one of Mel's daughters came in, along with Sadie, Mel's granddaughter. Six-year-old Sadie wore a blue skirt and smiled at me, shy and gap-toothed, between licks on a green Popsicle. A children's book has been written about Mel's property which features Sadie as "Sophie" and Mel as "Milton." In the book, the little girl spends part of a summer learning forestry from her grandfather. Mel's woods, the book reads, "are one of the finest sights in the country. In October, the leaves of the hardwoods turn copper, orange, red, yellow and purple, with pines and spruce towering green above."

When I returned out West, I found that one of the Montana landowners I met had suddenly died. I didn't know Shirley Bowdish well, but I was struck by the finality of her death—the fact that I couldn't ever see or talk to her again. The last time I saw her I remember she was wearing a pink cotton sweatshirt embroidered with ducks. She, her husband, John, and I had shared a dinner of stew and fresh bread after they had shown me their woodlot.

Unfortunately, many of the landowners I interviewed for this book are in the last years of their lives. They have seen logging trucks that used to fit only two big trees on their backs now carry dozens of thin trees. They have

Ames's mixed hardwood-softwood forest with stumps from a recent harvest

seen satellite dishes polka dot and roads candy stripe the deep country. After Shirley Bowdish's death, I wanted to make sure that I had gathered the strings of these landowners' lives—that I had learned and passed along what was important.

The landowners in this book vary in many ways, including the type of forest they have and the amount of income they earn from harvesting. But for all of them their forestland is one of the most important things in their lives. "I'm gettin' so I don't remember people's names and things," Leo Goebel said, "but I still remember all my trees."

These landowners recognize that careless management—too little or too much logging—could leave their forests burned down, or empty of everything but a few weak trees. They have all taken concrete measures to reduce the impact of their logging: carrying out an inventory and writing a management plan, finding a knowledgeable forester and a skilled logger, developing a plan for the future of their forest. Their goal in the end is to have a beautiful forest, a place neither too dense nor too barren, with rich soil and a variety of trees, plants, and wildlife. You may not be able to invest as much money or time as these landowners and walk your land so much that you give a nickname to each big tree. But I hope that you now search, as these landowners did, for an honest approach to resource use and conservation. Try to conserve and enhance your forest's beauty—the vigor of its trees, the delicacy of its flowers, the mysteries of its wildlife.

APPENDIX I

FINDING HELP:
Public and Private Assistance for Landowners

EXTENSION SERVICES
(useful for inexpensive forestry publications and workshops)

All of the extension services in the region provide quality, factual information to help you manage your forestland. Montana's workshop program, in particular, has been a model for others throughout the country. Contact your local extension service for a calendar of workshops and a listing of publications:

Colorado Extension Forestry
1 Administration Building
Colorado State University
Fort Collins, CO 80523-4040
(970) 491-6281
www.ext.colostate.edu/menunatr.html

Idaho Extension Forestry
College of Natural Resources
P.O. Box 441140
University of Idaho
Moscow, ID 83844-1140
(208) 885-6356
www.ets.uidaho.edu/extforest

Montana Extension Forestry
32 Campus Drive
University of Montana
Missoula, MT 59812
(406) 243-2773

Oregon Extension Forestry
119 Peavy Hall
Oregon State University
Corvallis, OR 97331
(541) 737-3700
www.osu.orst.edu/extension

Utah Extension Forestry
Utah State University
Logan, UT 84322
(435) 797-0560
http://extension.usu.edu/coop/natres/forests/index.htm

Washington Extension Forestry
Department of Natural Resources
Washington State University
Pullman, WA 99164
(509) 335-2963
http://ext.nrs.wsu.edu

STATE FORESTERS
(useful for forestry consulting and federal cost-share assistance)

The National Association of State Foresters is the umbrella organization for all state forestry branches. The Association works with the U.S. Forest Service's Cooperative Forestry to promote a number of programs for landowners, including the Forest Legacy Program, which helps large landowners with estate management and the Forest Land Enhancement Program which provides cost-share funding (typically 25 to 50 percent) for otherwise costly forest activities such as precommercial tree thinning and prescribed burning. State foresters are most helpful in providing advice at the beginning stages of writing a forest management plan or devising a timber sale. State foresters can provide you with some technical assistance and provide a list of foresters and loggers. They can also connect you with state and federal wildlife agencies (such as the U.S. Fish and Wildlife Service if you have endangered species) and help you get involved with various landowner assistance programs.

USFS Cooperative Forestry
P.O. Box 96090
Washington, D.C. 20060-6090
(202) 205-1389
www.fs.fed.us/spf/coop/

Intermountain Region
William Boettcher
Director of State and Private Forestry
USDA Forest Service
P.O. Box 7669
Missoula, MT 59807
(406) 329-3280 in Montana
wboettcher@fs.fed.us

National Association of State Foresters
Hall of the States
444 N. Capitol St., Suite 540
Washington, D.C. 20001
(202) 624-5415
www.stateforesters.org

Colorado
James E. Hubbard
CO State Forest Service
Colorado State university
203 Forestry Bldg.
Fort Collins, CO 80523
(970) 491-6303
jhubbard@lamar.colostate.edu

Idaho
Winston A. Wiggins
Idaho Department of Lands
954 West Jefferson Street
P.O. Box 83720
Boise, ID 83720
(208) 769-1577
wwiggins@idl.state.id.us

Montana
Tony Liane (acting)
DNRC — Forestry Division
2705 Spurgin Road
Missoula, MT 59804
(406) 542-4300
suclark@state.mt.us

Oregon
James E. Brown
Oregon Department of Forestry
2600 State Street
Salem, OR 97310
(503) 945-7211
jbrown@odf.state.or.us

Utah
A. Joel Frandsen
Department of Natural Resources
1594 W. North Temple, Suite 3520
Salt lake City, UT 84114-5703
(801) 538-5540
joelfrandsen@utah.gov

Washington
Pat McElroy
Department of Natural Resources
1111 Washington Street
Olympia, WA 98504-7001
(360) 902-1603
pat.mcelroy@wadnr.gov

Wyoming
Tom Ostermann
Wyoming State Forestry Division
1100 West 22nd Street
Cheyenne, WY 82002
(307) 777-7586
toster@state.wy.us

LANDOWNER ORGANIZATIONS
(useful for peer contacts, educational seminars, and forestry tours; the private version of extension services)

The American Tree Farm System is the nation's largest forest landowner program. The organizations' state conventions bring together a variety of experienced landowners, industry foresters, forestry consultants, and government representatives you can learn from. For your property to become a certified Tree Farm, you should contact:

The American Tree Farm System
1111 19th St. NW, Suite 780
Washington, D.C. 20036
1-888-889-4466
www.treefarmsystem.org

The National Woodland Owners Association (NWOA) lobbies Congress on behalf of landowners and has a number of affiliated state branches that sponsor educational workshops on topics like forest tax reduction and insect and disease control. You can join one of the state branches for a small annual fee.

National Woodland Owners Assoc.
374 Maple Ave. E., Suite 310
Vienna, VA 22180
1-800-GRO-TREE
www.woodlandowners.org

Colorado Forestry Association
John Oram, Director & Editor
5130 Meade Street
Denver, CO 80221
(303) 477-0552
jf-oram@msn.com

Idaho Forest Owners Assoc.
Kennon McClintock, President
HCR 62 Box 52
Moyie Springs, ID 83845
(208) 267-7064

Montana Forest Owners Assoc.
Thornton Liechty, President
17975 Ryans Lane
Evaro, MT 59802
(406) 726-3787
arl3787@blackfoot.net

Oregon Small Woodlands Assoc.
Mike Gaudern, Exec. Director
and Editor
1775 32nd Place NE, Suite C
Salem, OR 97303
(503) 588-1813
sgennett@oswa.org

Utah Woodland Owners Council
Richard Oldroyd, President
2829 Sleep Hollow Drive
Salt lake City, UT 84117
(801) 277-1615

Washington Small Woodlands Assoc.
Nels Hanson, Assoc. Exec. Director
110 West 26th Avenue
Olympia, WA 98507
(360) 459-0984
nelswh@home.com

PRIVATE FORESTRY CONSULTANTS
(useful for timber sales and ongoing forest management; the private version of state foresters)

The Society of American Foresters (SAF) is a national educational organization. It was founded in 1900 by the country's first national forester, Gifford Pinchot. The organization publishes scientific journals on forestry *(The Journal of Forestry, Forest Science,* and *The Western Journal of Applied Forestry)* and has a listing of forestry consultants.

Society of American Foresters
5400 Grosvenor Lane
Bethesda, MD 20814
(301) 897-8720
www.safnet.org

The Forest Stewards Guild, founded in 1997, is an organization of natural resource professionals and consultant foresters committed to "ecologically responsible resource management that sustains the entire forest across the landscape." The organization has a database of consultant foresters, restoration ecologists, loggers, and others who practice sustainable forestry on the ground.

Forest Stewards Guild
P.O. Box 8309
Santa Fe, NM 87504-8309
(505) 983-3887
www.foreststewardsguild.org

NONPROFIT ORGANIZATIONS
(useful for conservation easements and timber certification)

The total acreage in conservation easements has more than quadrupled in the West over the last decade. The Nature Conservancy, the nation's largest easement organization, focuses on lands with unique scientific value. The Land Trust Alliance is an umbrella organization for a network of regional organizations (such as the Montana Land Reliance) that establish easements on all types of forests.

The Land Trust Alliance
1331 H. Street NW, Suite 400
Washington, D.C. 20005-4734
(202) 638-4725
www.lta.org

The Nature Conservancy
4245 N. Fairfax Dr., Suite 100
Arlington, VA 22203-1606
(800) 628-6860
www.tnc.org

The Nature Conservancy
Conservation Forestry Program
Clinch Valley Chapter
146 East Main Street
Abingdon, VA 24210
(276) 676-2209

The Trust for Public Land
116 New Montgomery St., 4th Floor
San Francisco, CA 94105
(415) 495-4014
www.tpl.org

SmartWood, based in Vermont, has a national timber certification program for landowners with rigorous ecological standards. The Forest Stewardship Council is the parent organization for SmartWood's certification program.

SmartWood
Goodwin-Baker Building
65 Millet Street, Suite 201
Richmond, VT 05477
(802) 434-5491
www.smartwood.org

Forest Stewardship Council (U.S.)
1155 30th Street NW, Suite 300
Washington, D.C. 20007
1 (877) 372-5646
www.fscus.org

The Certified Forest Products Council has a database of certified forest products for you to consider when you need to buy wood products. These products, ranging from guitars to flooring, are often more expensive than uncertified goods, but are harvested sustainably.

Certified Forest Products Council
721 NW 9th Avenue, Suite 300
Portland, OR 97209
(503) 224-2205
www.certifiedwood.org

Innovative groups in the Northeast and Northwest also promote sustainable forestry and may be able to help landowners in these regions find a good forester,
develop a conservation easement, or get certified.

Institute for Sustainable Forestry
P.O. Box 1580
Redway, CA 95560
(707) 247-1101
www.isf-sw.org/aboutISF.htm

Pacific Forest Trust
416 Aviation Blvd., Suite A
Santa Rosa, CA 95403
(707) 578-9950
www.pacificforest.org

New England Forestry Foundation
238 Old Dunstable Rd.
Groton, MA 01450
(978) 448-8380 x109
www.neforestry.org

Society for the Protection of
New Hampshire Forests
54 Portsmouth St.
Concord, NH 03301
(603) 224-9945
www.spnhf.org

Landowner cooperatives are still young and unproven, but they offer new possibilities for forest owners to work together. In addition to the co-op and landowner
organizations below, you may want to contact two organizations that provide support for co-ops—the Community Forestry Resource Center in Minnesota
(www.forestrycenter.org) and the University of Wisconsin Center for Cooperatives (www.wisc.edu/uwcc).

Sustainable Woods Cooperative
P.O. Box 307
Lone Rock, WI 53566
(608) 583-7100
www.sustainablewoods.com

Vermont Family Forests
P.O. Box 254
Bristol, VT 05443
(802) 453-7728
www.familyforests.org

Taxes are often particularly befuddling to landowners. Check www.timbertax.org
for assistance.

In addition to the resources listed here, the Bibliography has a number of useful
books arranged by subject.

APPENDIX II

Sample Timber Harvest Contract

A good timber harvest contract will help protect you from problems during and after the logging. Even if you prefer to do business on a handshake, it's important to write a contract for timber sales to protect your land. You can start by referring to this sample contract, then contacting your attorney.

CAPTION AND RECITAL

The following clauses set the background for the contract. The "caption" introduces the parties in the contract and the "recital" states the parties' intentions.

This is an agreement made on March 15, 1984, between Mr. and Mrs. Jim Johnson, P.O. Box 000, Coram, MT 59913, fictitious entities hereinafter referred to as the "Johnsons" or "Sellers," and Mountain Lumber Co., P.O. Box 000, Kalispell, MT 59882, a fictitious Montana Corporation hereinafter referred to as "Mountain Lumber" or "Buyer."

Whereas, the Johnsons own certain lands upon which merchantable timber is standing; Whereas, the Johnsons desire and have the right to sell the same; Whereas, Mountain Lumber is willing to purchase the same upon certain conditions. Now, therefore, in consideration of the following mutual covenants, it is agreed as follows:

DESCRIPTION OF THE TIMBER SALE

This section describes the wheres and whats of harvesting. Basically, Mountain Lumber will harvest trees that the landowners have already marked, perhaps in consultation with a professional forester. Sometimes landowners will mark trees at both breast height and at the base so they can tell from the stumps whether the logger has taken the right trees. On the other extreme, landowners might not mark any trees at all for loggers they have already worked with and only lay out rough guidelines for the type of trees they want taken.

In this contract, special environmental provisions require skidding over snow to reduce soil damage; washing of machinery in the spring, summer, and fall to prevent dispersal of noxious weed seeds; and judicious harvesting to retain wild-life trees. The Streamside Management Zone Law further restricts harvesting within 50 feet of perennial streams. Note the severe penalty – three times the wood value – for removing unmarked trees or not removing marked trees.

Timber Involved. The Johnsons agree to sell merchantable timber suitable for sawlogs from the following property, hereinafter referred to as "the Property": S1/2 Section 30 and SW1/4, SW1/4 Section 31; T30N, R21W (approximately 30 acres).

Access. Sellers grant to Mountain Lumber the exclusive right-of-way across the Property during the life of this contract for the removal of timber and for constructing and utilizing necessary improvements. Mountain Lumber will obtain permission for crossing adjacent land, if necessary. Mountain Lumber will repair at its expense damage incurred to existing roads, bridges, fences, or other improvements.

• Construction of roads and operation of machinery will follow the Streamside Management Zone Law and Best Management Practices to protect water quality. Existing roads will be used whenever possible, and new road locations must be mutually agreed upon, in good faith, by Mountain Lumber and the Johnsons. Felling, skidding, and slash-piling will only be done when the ground is covered with at least six inches of snow.

• Mountain Lumber will purchase and apply native grass and forb seed to skid trails and landings in the spring after harvesting, if requested. Mountain Lumber will also wash its equipment prior to entering the site in the spring, if requested, to prevent infestations of noxious weeds.

Treatment. The Johnsons and Mountain Lumber agree that the property will be selectively thinned unless otherwise arranged. This logging operation will remove mature and immature timber that has been painted with a blue slash by the landowner at breast height and at the stump. Stump height will normally not exceed 8 inches. Mountain Lumber will not harvest in areas it deems unsafe because of topography or stand conditions.

• Wildlife habitat will be preserved by: leaving all dead, standing trees (snags); creating raptor perches, 15 to 20 feet tall, averaging 1 per acre; and creating 3- to 4-foot-tall stumps for small mammal perches, averaging 0.5 per acre. For aesthetics, unharvested areas 25 feet wide will be retained on each side of county, state, and federal roads.

Merchantability. Specifications for sawlog merchantability for all species include: minimum sawlog length of 16 feet, minimum dbh of 9 inches, minimum top diameter inside bark of 6 inches, and minimum soundness of 50 percent. Pulpwood or posts and poles, trees too small for sawlogs, will be removed at the Sellers' discretion. A penalty of three times the stumpage rate will be assessed for removing or carelessly damaging any trees not designated for cutting or for failing to remove designated trees.

Timeline. Mountain Lumber may access the property anytime after January 15, 1984, and up to May 30, 1985. Unless otherwise arranged, all timber remaining on the site after the termination date will become the property of the Johnsons.

SLASH MANAGEMENT AND FIRE CONTROL

Montana, like many other western states, requires a cash deposit to ensure that logging residue is burned to prevent catastrophic fires. State employees inspect the site, then refund the deposit if the disposal is satisfactory. Note in the section below that Mountain Lumber must separate merchantable material from the slash so the Johnsons can recover some of that wood to use it for firewood or fence posts rather than burning it as slash.

Slash Disposal. Mountain Lumber agrees to deposit $6.75 per thousand board feet (mbf) to the State of Montana for Fire Hazard Reduction. Tree-length skidding to landings will be used in most areas to minimize soil disturbance. Merchantable post and pole, firewood, and pulp material will be decked separately from the other slash material. Buyer must pile all slash by June 30, 1985, and dispose of the slash by December 1, 1986. After state officials accept the slash disposal, Buyer will receive the returned deposit.

Fire Control. Mountain Lumber will prevent and suppress forest fires on or near the sale area during its harvesting operations. Mountain Lumber will accept responsibility for fires that occur because of their actions or the actions of their contractors, but not fires started by natural causes or by a third party.

Regeneration. All regeneration will come from natural means. Buyer is not responsible for replanting.

SCALING TREES

This section describes the rules for scaling, or measuring, trees to determine their volume and price.

Scaling. All logs cut from the property shall be scaled in accordance with the rules of the U.S. Forest Service scaling handbook FSH 2409.11 using the Scribner Decimal C Scale Rule. All logs delivered to the mill will be 100 percent scaled at the Mountain Lumber yard in Missoula, Montana. The Johnsons are welcome to scale logs themselves at any time. If a discrepancy is found, a third party, acceptable to both Buyer and Sellers, will submit his own scale. This decision shall be final and binding, and the costs of the check scaling will be shared equally by both parties.

PAYMENTS

After the scaling rules are set, the financial details are laid out: the prices for timber removed and the time frame for the payments. Load tickets, which show the volume of wood brought to the mill, provide evidence that the loggers paid the landowners fairly.

Stumpage Rate. Payment will be based on net scale figures. Mountain Lumber agrees to pay the Johnsons $200.00 per mbf (Scribner) for all merchantable timber cut and removed. Buyer agrees to send consecutive load tickets and cash payments to the Sellers on the 15th and 30th of each month, within 15 days of receiving the logs at the mill. Title for the timber shall pass to Mountain Lumber upon payment to the Johnsons.

WARRANTY

This paragraph clarifies the ownership of the land and timber and relieves Mountain Lumber from property disputes, including payment of attorney's fees. Legal fees are generally not covered, even under winning court settlements, unless clarified in the contract. Mountain Lumber, in turn, protects the Johnsons from liens, or claims that Mountain Lumber employees might make against the Johnsons in a lawsuit.

Seller's Warranty. The Johnsons warrant that they own the above described property and have the lawful right to sell its timber, free and clear. The Sellers hold Mountain Lumber harmless from all costs, including attorney's fees, arising from third party claims to the timber. The sellers will not suffer any labor liens or other encumbrances levied against the property.

BEGINNING OF LIABLILTY

Mountain Lumber agrees to carry insurance against accidents that may take place during the harvesting operations. Mountain Lumber also assumes independent contractor status, which protects the Johnsons from workers' compensation claims if an employee is injured on the job, and relieves the Johnsons from paying employee taxes.

Liability Insurance. At its own cost, Mountain Lumber will maintain insurance policies against bodily injury or property damage, including events involving motorized vehicles, at a minimum of $50,000 for property damages; $500,000 for injury or death to any person; and $1,000,000 for injury or death to multiple people arising from a single incident, including fire. In addition to liability insurance, Mountain Lumber holds workers' compensation insurance for its employees.

Contractor Status. Mountain Lumber will act as an independent contractor and pay all charges incurred in its logging operations.

ACTS OF GOD

This "Act of God" clause appears in most contracts. Note that a reduction in timber prices is included here along with natural disasters.

Force Majeure. Should either party be prevented from upholding this agreement by causes beyond their control, such as, but not limited to, strikes, flood, fire, or a marked reduction in timber prices, this agreement shall be suspended without cost to either party until normal operations can resume.

PERFORMANCE BOND AND HANDLING DISPUTES

The Johnsons will enforce the contract by demanding a performance bond and regularly monitoring logging. If disputes arise, both Mountain Lumber and the Johnsons will first try to resolve their problems out of court.

Performance Compliance. When this contract is signed, a bond of 15 percent of the timber value will be paid to the Sellers, to be held for the life of the contract to ensure faithful compliance.

Default. In the case of default, the defaulting party shall receive written notice stating the manner of default. If the default is not corrected within 10 days of receipt of the notice, all logging operations will be halted and the contract will be terminated. Termination will not relieve either party of damage payments, including repayment of compliance bonds, with interest at 10 percent per annum.

Arbitration. In the case of disputes, an arbitration board will be established with three members, one selected by the Sellers, one by the Buyer, and a third selected by the first two members. The board must reach an agreement within 30 days and their decision will be binding on both parties, as further set forth in section 27-5-111 et seq. Montana Code Annotated. The costs of arbitration will be shared.

FUTURE OF CONTRACT

This section binds anyone who buys or inherits the land to the contract during its duration. Mountain Lumber may also arrange for another company, an "assign," to carry out the contract. However, Mountain Lumber remains responsible for any violations incurred by an assign.

Inurement. This agreement shall be binding upon the successors, heirs, and assigns of the respective parties.

IN WITNESS WHEREOF the parties have signed this agreement on the date set out above.

APPENDIX III

PREPARING A CONSERVATION EASEMENT

Below are examples of the steps you can take to draw up a pre-easement resource document and a conservation easement, based on materials provided by the Montana Land Reliance. Planning an easement should prompt you to ask of your own land:

What could I accomplish with an easement?

What is the history of my land?

What is surrounding my land?

Which plant and animal resources do I have on my property, and which am I particularly concerned about protecting?

Which rights am I willing to give up to reduce my estate taxes and protect my land and which do I need to retain?

How am I going to define the parameters for timber harvesting?

THE PRE-EASEMENT RESOURCE DOCUMENT

Before writing an easement, a conservation organization will draw up a pre-easement resource document, which identifies and describes the following:

Conservation purpose

"The Johnson property is located in Lodgepole County in northwestern Montana in Sections 30&31, Township 30 North, Range 21 West, near the town of Lumberton. The conservation purposes of the easement, which fulfill the conservation purposes defined by Treasury Regulations Sec. 1.170A-14 (d) (1), include:

protection of relatively natural habitat for fish, wildlife, and plant communities; and preservation of forested open space and potential wildlife corridors in an area that is being cleared for pasture, timber or housing."

Land use history

". . . The Northern Pacific Railroad acquired the land soon after 1887 when the Dawes Act was passed, opening reservation land to non-Indian use, then Northern Pacific offered the land for public sale in 1938. . . . The Johnsons initially purchased 80 acres (currently the eastern half of their property) in 1967 from Mr. Jack Green. In 1971, the Johnsons purchased the remaining 80 acres, or the western half of their property, from Mrs. Leonora Dolton.

"The western 80 acres of mixed species had been logged in 1964, leaving a tremendous amount of debris that the Johnsons piled and burned. In 1973, the Johnsons began to plant Scotch pine for Christmas trees on a cutover area in the northwest corner of their property . . . This site is currently identified in the conservation easement as a potential building site for the future due in part to its past disturbances. The eastern 80 acres, not covered under this easement, are largely lodgepole pine that has not been disturbed since a fire roared through the valley 110 years ago."

Current land management

"The Johnsons have been harvesting lodgepole pine in one-quarter-acre group selections on the eastern half of their property for the last 10 years. The Johnsons wish to open the dense lodgepole pine stand for wildlife travel corridors and for planting other tree species that were historically more common in the area. The Johnsons also thin to a 10-foot spacing in the mixed-conifer forest on the western half of their property in order to restore the logged-over stand to three or more age classes of Douglas-fir, Engelmann spruce, subalpine fir, and western larch."

Adjacent land use

"All of the surrounding land is under private ownership. Most of the adjacent land is in similar condition to the Johnson property except to the north where the owner clearcut considerable acreage 5 years ago to convert it to pasture. . . . The Johnsons have since experienced windthrow on this edge of their property, as well as increased exotic species invasion."

Existing and planned structures, fences, roads and utilities

"A cabin that serves as the Johnson residence is located in approximately the center of the property. It was a two-story house that the Johnsons moved from the Ninemile area and converted to a one-story building with a loft. An outhouse and a woodshed are behind the residence to the south and a small horse corral, tractor shed, and old foundation that was built by previous owners are to the east of the cabin along the road. The Johnsons intend to complete a barn at the old foundation site if a second building is not constructed at the potential building site in the northwest corner of the property.

". . .Fencing on the Johnson property includes jack-leg fencing along the northern and western boundaries and along the western portion of the southern boundary. The Johnsons intend to build similar fencing along the eastern boundary and eastern half of the southern boundary to delineate the entire perimeter of the property.

"The cabin and surrounding structures are accessible from one major dirt road that enters the property in the northwest corner and runs approximately through the center of it. This road is gated on the northwestern end."

Vegetation

"Overall, six habitat types were identified on the Johnson' property:
1. subalpine fir/queencup beadlily
2. spruce/red-osier dogwood
3. Drummond willow
4. subalpine fir/bluejoint reedgrass
5. beaked sedge
6. common cat-tail

"The most widely distributed trees on the property are:
1. grand fir
2. subalpine fir
3. Engelmann spruce
4. lodgepole pine
5. quaking aspen
6. black cottonwood
7. Douglas-fir

"The most widely distributed shrubs are:
1. mountain alder
2. western serviceberry
3. kinnikinnick
4. Oregon grape
5. common juniper
6. twinflower
7. Wood's rose
8. Canada buffaloberry
9. shiny-leaf spiraea
10. snowberry
11. dwarf huckleberry

"The most widely distributed native forbs are:
1. yarrow
2. pearly everlasting
3. field pussytoes
4. mountain arnica
5. smooth aster
6. lady's thimble
7. prince's pine
8. queencup beadlily
9. bunchberry dogwood

10. blueleaf strawberry

11. northern bedstraw

12. balsam groundsel

13. starry Solomon plume

14. Canada goldenrod

15. western meadowrue

16. American vetch

"The most widely distributed ferns and fern allies are:

1. common scouring rush

"The most widely distributed native grasses, sedges, and rushes are:

1. bluejoint reedgrass

2. pinegrass

3. sedge species

4. blue wildrye

"The most widely distributed exotic forbs are:

1. oxeye daisy

2. Canada thistle

"The most widely distributed exotic grasses are:

1. common dandelion

2. redtop

3. timothy

"The mammals observed by the landowner include:

1. black bear

2. grizzly bear

3. mountain lion

4. coyote

5. elk

6. moose

7. white-tailed deer

8. badger

9. marten

10. weasel

11. chipmunk

12. ground squirrel

13. red-backed vole

14. deer mouse

15. shrew

16. beaver

17. mule deer

18. muskrat

"The birds observed by the landowner include:
1. ruffed grouse
2. pileated woodpecker
3. downy woodpecker
4. hairy woodpecker
5. great horned owl
6. barred owl
7. gray jay
8. Steller's jay
9. common raven
10. wren
11. cedar waxwing
12. hermit thrush
13. rufous hummingbird
14. black-chinned hummingbird
15. flicker
16. mountain chickadee
17. robin
18. nuthatch
19. junco"

THE CONSERVATION EASEMENT

After the resource document is reviewed by the landowners, a conservation organization will draw up an actual conservation easement, also with a number of components:

Enforceability and legal validation of easement

"Upon prior notice to the landowner, the Montana Land Reliance (MLR) reserves the right to enter upon the property in an unobtrusive manner (approximately once per year) to monitor compliance with the terms of this easement. But this easement does not grant MLR, nor the public, any other right to enter upon the property. MLR also reserves the right to enjoin any activity on, or use of, the property which is inconsistent with the purpose of this easement and to enforce reasonable restoration of the property damaged by such activity or use."

Costs and taxes

"The landowners shall bear all costs and liabilities of any kind related to the ownership, operation, upkeep, and maintenance of the property, including responsibility for the control of noxious weeds in accordance with Montana law. The landowners shall pay any and all taxes, assessments, fees, and charges levied on the property. Any tax or assessment on this easement, however, shall be paid by MLR."

Acceptable activities

"It is the purpose of this easement to assure that the conservation values will be maintained forever and to prevent any use of, or activity on, the property that will significantly impair those values. The landowner understands that this easement will limit the use of the property to such activities as are consistent with that purpose (including limited livestock uses; hunting, fishing, and other recreational uses; bed and breakfast business; and selective timber harvesting consistent with the terms hereof). The landowner may:

"Raise, graze, stable, and use for work or recreational purposes a maximum of four (4) horses or mules on the property. If the area bordering Buck Creek is used for livestock grazing, the landowners shall establish a fenced buffer zone around that riparian area.

"Use the property for hunting and fishing.

"Develop water resources on the property, provided that such development maintains the current natural conditions of water courses and wetlands.

"Maintain, repair, remodel, and make limited additions to any existing structures, roads, fences, pipelines, bridges, or culverts, or replace them with similar structures in the same general locations if they are destroyed. Also, the right to construct one new building in one of two identified locations on the property.

"Use biological control agents and agrichemicals including fertilizers, pesticides, herbicides, insecticides, and rodenticides in those amounts necessary to accomplish reasonable forestry and residential objectives.

"Rent the cabin on the property. The landowners can also develop a forest nursery, not larger than five (5) acres for the propagation, harvest, and sale of native conifer trees, shrubs, and grasses.

"Sell, exchange, or gift the property as one (1) parcel. The landowners shall furnish MLR a copy of any document or conveyance within thirty (30) days of the execution of the document.

"Selectively harvest timber (including group selections) in accordance with all applicable state and federal forestry laws, practices, guidelines, and regulations, including Montana Forestry Best Management Practices (BMPs) and the Streamside Management Zone Law. Noncommercial harvesting, where the trees are used or disposed of on the property, is a reserved right if being done to remove dead or diseased trees or to create noncommercial firewood or posts and poles. However, any commercial harvest or thinning, including those to control insect or disease infestation, will require a timber harvest plan, approved by a state-qualified extension forester or consultant, which includes information required by the MLR. The MLR and landowner will mutually determine the completeness of the timber harvest plan and its adherence to the intentions of this easement before the harvest is begun. Once begun, the commercial harvest must minimize impacts on the watershed, water quality, wildlife habitat, and forest aesthetics. In the event of a severe fire, windstorm or insect damage, the land shall be allowed to return to a forest ecosystem.

Prohibited activities

"In accordance with the easement restrictions, the landowner shall not:

"Divide or subdivide the property.

"Explore or extract any materials for commercial purposes by mining (e.g. oil, gas, minerals, gravel, peat, rock, etc.).

"Establish any commercial or industrial facilities (e.g. guest ranching, outfitting, retail sales, campground, motel, etc.).

"Dump non-compostable refuse.

"Change, disturb, alter, or impair any watercourse or wetland.

"Construct any additional roads except for eminent domain or temporary roads for selective timber harvesting.

"Construct a commercial feedlot.

"Install utility structures, lines, conduits, cables, wires, or pipelines on the property, except in connection with existing and permitted structures. A buried utility right-of-way is allowed only on the northern boundary of the property.

"Construct, maintain or erect any billboards.

"Place, use or maintain any trailer, mobile home, or other movable living unit on more than a temporary basis.

"Raise or confine for commercial purposes game animals, native or exotic fish, game birds, furbearers, farm animals, or other native or exotic animals.

"Raise pigs, sheep, goats, chickens, cattle, bees, or other agricultural animals.

"Use the property for farming, ranching, or other agricultural activities, besides a limited garden and restricted timber harvesting.

"Store, place, or dispose of animal or human food in areas accessible to wildlife."

GLOSSARY

allelopathic. A plant that inhibits the growth of another plant species by emitting chemical substances.

annulus ring. A thin membrane encircling the stem on some mushrooms.

bucking. Cutting trees into merchantable lengths.

certification. A third-party review of forest management and timber harvesting to ensure compliance with environmental and social standards.

choker. A clasp that closes around the butts of trees to cinch them as they are dragged behind a cable skidder.

commercial thinning. Removal of merchantable timber in order to improve forest structure or encourage growth of remaining trees.

crop trees. Particularly valuable trees, based on their straightness and species, which are often fattened up by thinning away surrounding trees.

dimensional lumber. Wood for building frame construction.

epicormic branching. Branching from dormant buds in a trunk—a particular problem among deciduous trees that receive too much light at once.

feller-buncher. A machine commonly used for harvesting that has arms that clamp around a trunk and a sawblade that comes from the bottom to snip the tree off the stump.

forester. A professional with an accredited forestry degree who helps a landowner develop a plan and hire personnel to carry out the landowner's forest management objectives.

forest health. The ability of a forest to resist and quickly regrow after disturbances such as fires, windstorms, and insect and disease infestations.

harvester. A machine that bucks trees while still in the forest, rather than skidding trees numerous times from the forest to the landing.

inurement. A legal term for binding through use, such as contract terms binding successive parties.

jackpot burns. Burning wood in piles, as opposed to broadcast burning.

leave trees. Trees left in a forest after a harvest for their ecological or seed value.

old-growth forest. A forest defined variously by age (with trees over 150 to 200 years old), structure (with three or more age classes), or ecological process (with a mosaic of tree species on a small scale but a stable composition on a large scale through time).

over-aged trees. Trees that have so slowed in growth or become so susceptible to disease due to age that they should be harvested unless they have particular ecological value.

peeler. A large tree, generally greater than 16 inches in diameter, that can be sold for furniture veneer or plywood.

performance bond. A sum of money that a logger or other contractor deposits to insure that job performance is met according to contract.

precommercial thinning. Removal of unmerchantable timber in order to improve ecological structure or encourage growth of remaining trees.

saprophytic. Absorbing nourishment from dead or decaying matter.

scaling. Measuring board foot volume in trees, often for payment.

seral. Referring to the ecological or successional stage of a particular plant community. See **succession.**

skidding. Moving trees from the forest to a roadside landing where they can be bucked to length and trucked to the mill.

slash. Debris, such as limbs and tree tops, left after harvesting.

snags. Dead standing trees, often valuable as wildlife habitat.

stocking. A subjective estimate of forest density relative to growth.

succession. The natural process through which one plant community replaces another as an area matures ecologically or recovers from a disturbance.

sustainable forestry. Harvesting timber in a way that maintains the diversity, function, and structure of the forest.

BIBLIOGRAPHY BY SUBJECT

Fire and fire ecology

Agee J. *Fire Ecology of Pacific Northwest Forests.* Washington, D.C.: Island Press, 1993.

Arno, S. and S. Allison-Burnell. *Flames in Our Forest.* Washington, D.C.: Island Press, 2002.

Gruell, G. *Fire in Sierra Nevada Forests: A Photographic Interpretation of Ecological Change Since 1849.* Missoula, Mont.: Mountain Press, 2001.

Maclean, N. *Young Men and Fire.* Chicago: University of Chicago Press, 1993.

Pyne, S. and W. Cronin. *Fire in America.* Seattle: University of Washington Press, 1997.

Forestry

Best, C. and L. A. Wayburn. *America's Private Forests: Status and Stewardship.* Washington, D.C.: Island Press, 2001.

Dobbs, D. and R. Ober. *The Northern Forest.* White River Junction, Vt.: Chelsea Green Publishing, 1996.

Lansky, M. *Beyond the Beauty Strip.* Gardiner, Maine: Tilbury House, 1992.

Manning, R. *The Last Stand.* New York: Penguin Books, 1991.

Moore, B. *The Lochsa Story: Land Ethics in the Bitterroot Mountains.* Missoula, Mont.: Mountain Press, 1996.

Shetterly, S. H. *Shelterwood* (children's book). Illustrated by R. H. McCall. Gardiner, Maine: Tilbury House, 1999.

Forest ecology

Oliver, C. D. and B. C. Larson. *Forest Stand Dynamics.* New York: McGraw-Hill Publishing, 1990.

Smith, D., ed. *The Practice of Silviculture.* New York: John Wiley & Sons, 1996.

Vogt, K. A., et al., eds. *Forest Certification: Roots, Issues, Challenges, and Benefits.* Boca Raton, Fla.: CRC Press, 1999.

Forest economics and estate management

McEvoy, T. *Legal Aspects of Owning and Managing Woodlands.* Washington, D.C.: Island Press, 1998.

Small, S. *Preserving Family Lands: Essential Tax Strategies for the Landowner.* Washington, D.C.: Landowner Planning Center, 1992.

Small, S. *Preserving Family Lands, Book II: More Planning Strategies for the Future.* Washington, D.C.: Landowner Planning Center, 1997.

Small, S. *Preserving Family Lands, Book III: New Tax Rules and Srategies and a Checklist.* Washington, D.C.: Landowner Planning Center, 2002.

Forest health

Hagle, S. K., S. Tunnock, K. E. Gibson, and C. J. Gilligan. *The Field Guide to Diseases and Insect Pests of Idaho and Montana Forests.* Publication R1-89-54. Missoula, Mont.: USDA Forest Service, 1990.

Langston, N. *Forest Dreams, Forest Nightmares.* Seattle: University of Washington Press, 1995.

Little, C. *The Dying of the Trees.* New York: Penguin, 1997.

Sampson, R. N., D. L. Adams, and M. Enzer, eds. *Assessing Forest Ecosystem Health in the Inland West.* Binghamton, NY: Food Products Press, 1994.

Forest management for landowners

Beattie, M., et al. *Working with Your Woodland.* Hanover, N.H.: University Press of New England, 1993.

Bundy, P. *Finding the Forest.* Crosby, Minn.: Masconomo Forestry, 1999.

Camp, O. *The Forest Farmer's Handbook: A Guide to Natural Selection Forest Management.* Ashland, Oreg.: Sky River Press, 1984.

Craighead, C. *Who Ate the Backyard? Living with Wildlife on Private Land.* Moose, Wyo.: Grand Teton Natural History Association, 1997.

Fazio, J. *The Woodland Steward.* Moscow, Idaho: The Woodland Press, 1985.

Hilts, S. and P. Mitchell. *The Woodlot Management Handbook.* Toronto: Firefly Books, 1999.

Loomis, R. with M. Wilkinson. *Wildwood: A Forest for the Future.* Gabriola, B.C., Canada: Reflections Press, 1995.

Walker, L. *Farming the Small Forest: A Guide for the Landowner.* San Francisco: Miller Freeman, 1988.

Sustainable forestry

Aplet, G., et al. *Defining Sustainable Forestry.* Washington, D.C.: Island Press, 1993.

Davis, T. *Sustaining the Forest, the People and the Spirit: The Story of the Monominee Indian Tribe.* New York: State University of New York Press, 2000.

Drengson, A. R. and D. Taylor, eds. *Ecoforestry: The Art and Science of Sustainable Forest Use.* Gabriola, B.C., Canada: New Society Publishers, 1998.

Kohm, K. A. and J. F. Franklin, eds. *Creating a Forestry for the 21st Century: The Science of Ecosystem Management.* Washington, D.C.: Island Press, 1997.

Maser, C. *Sustainable Forestry: Philosophy, Science, and Economics.* Boca Raton, Fla.: St. Lucie Press, 1997.

INDEX

Page numbers in italics refer to photographs or illustrations.

MARCIE JOHNSON PHOTO

ABOUT THE AUTHOR

A native Montanan, Bryan Foster first became interested in forests as a boy exploring woodlands on his father's farm. Foster holds a master's degree in forestry from Yale University and has written for a number of publications, including *High Country News*. He currently lives in Santa Fe, New Mexico.

We encourage you to patronize your local bookstore. Most stores will order any title they do not stock. You may also order directly from Mountain Press, using the order form provided below or by calling our toll-free, 24-hour number and using your VISA, MasterCard, Discover or American Express.

Other books of interest:

_____DEADFALL:
Generations of Logging in the Pacific Northwest   14.00

_____FIRE IN SIERRA NEVADA FORESTS:
*A Photographic Interpretation of Ecological
Change since 1849*   20.00

_____FORTY YEARS A FORESTER: 1903 – 1943   paper   16.00
     cloth   30.00

_____THE LOCHSA STORY:
Land Ethics in the Bitterroot Mountains   paper   20.00
     cloth   36.00

_____SOUTH OF SEATTLE:
Notes on Life in the Northwest Woods   10.00

_____WILD LOGGING:
*A Guide to Environmentally and Economically
Sustainable Forestry*   16.00

Please include $3.00 per order to cover postage and handling.

Send the books marked above. I enclose $_____

Name _____

Address _____

City/State/Zip _____

☐ Payment enclosed (check or money order in U.S. funds)

Bill my: ☐ VISA   ☐ MasterCard ☐ Discover ☐ American Express

Card No. _____ Expiration Date:_____

Signature _____

MOUNTAIN PRESS PUBLISHING COMPANY
P.O. Box 2399 • Missoula, MT 59806 • Order Toll-Free 1-800-234-5308
E-mail: info@mtnpress.com • Web: www.mountain-press.com